ENVIRONMENTAL
JUSTICE

A Reference Handbook

Other Titles in ABC-CLIO's
CONTEMPORARY
WORLD ISSUES
Series

Books in the Contemporary World Issues series address vital issues in today's society such as terrorism, sexual harassment, homelessness, AIDS, gambling, animal rights, and air pollution. Written by professional writers, scholars, and nonacademic experts, these books are authoritative, clearly written, up-to-date, and objective. They provide a good starting point for research by high school and college students, scholars, and general readers, as well as by legislators, businesspeople, activists, and others.

Each book, carefully organized and easy to use, contains an overview of the subject; a detailed chronology; biographical sketches; facts and data and/or documents and other primary-source material; a directory of organizations and agencies; annotated lists of print and nonprint resources; a glossary; and an index.

Readers of books in the Contemporary World Issues series will find the information they need in order to better understand the social, political, environmental, and economic issues facing the world today.

ENVIRONMENTAL
JUSTICE

A Reference Handbook

David E. Newton

CONTEMPORARY WORLD ISSUES

ABC-CLIO

Santa Barbara, California
Denver, Colorado
Oxford, England

Copyright © 1996 by Instructional Horizons, Inc.

Library of Congress Cataloging-in-Publication Data

Newton, David E.
    Environmental justice : a reference handbook / David E. Newton.
        p.   cm. — (Contemporary world issues)
    Includes bibliographical references and index.
        1. Distributive justice. 2. Environmental policy—Social aspects.
    3. Civil rights. I. Title. II. Series.
    JC578.N485   1996   363.7—dc20

ISBN 0-87436-848-0 (alk. paper)   96-26539
                                                                          CIP

01 00 99 98 97 96 95   10 9 8 7 6 5 4 3 2 1   (cloth)

ABC-CLIO, Inc.
130 Cremona Drive, P.O. Box 1911
Santa Barbara, California 93116-1911

This book is printed on acid-free paper ∞ .
Manufactured in the United States of America

For Delores and Reg

Contents

Preface, xiii

1 The Road to Environmental
Justice, 1
Warren County, North Caro-
lina: 15 September 1982, 1
Environmental Inequities,
Environmental Racism, and
Environmental Justice, 2
Examples of Environmental
Inequities, 5
West Dallas, Texas, 6
Churchrock, New Mexico, 7
"Cancer Alley," 9
Pesticide Exposure, 11
Consumption of Toxic Fish
from the Detroit River, 12
Environmental Inequities on
an International Level, 13
Environmental Justice: Civil
Rights + Environmentalism,
15
The Modern Environmental
Movement, 16

The Civil Rights Movement, 18

Environmental Justice Comes of Age, 19
The Commission on Racial Justice of the United
Church of Christ, 20
The Michigan Conference, 22
Governmental Recognition, 22
First National People of Color Environmental
Leadership Summit, 23

Environmental Justice as a Social Movement, 24

Data and Statistics, 25
Race, Economic Status, and Environmental
Inequities, 26

Differences of Opinion, 29

The Origins of Environmental Inequities, 31
Progress and Environmental Degradation, 31
NIMBY and PIBBY, 32
Concerns about Environmental Issues among People
of Color, 33
Access to Tools of Protest, 34
Job Blackmail, 35
The Effects of Environmental Regulations, 36

Environmental Inequities: By Chance
or By Choice?, 39

The Legal Question of "Intent" in
Environmental Inequities, 44

Responding to Environmental Inequities, 45

Environmental Justice as an International
Issue, 47

The Future of Environmental Justice, 49

References, 51

2 Chronology, 57

3 Biographical Sketches, 67
Bunyan Bryant (1935–), 68
Robert Bullard, 69
César Chávez (1927–1993), 70
Benjamin F. Chavis (1948–), 71

Jay Feldman (1953–), 72
Deeohn Ferris (1953–), 73
Neftalí García Martínez (1943–), 74
Clarice E. Gaylord (1943–), 75
Juana Beatriz Gutierrez (1932–), 75
LaDonna Harris (1931–), 76
Pamela Tau Lee, 77
Les Leopold (1947–), 78
John Lewis (1940–), 79
Paul Mohai (1949–), 79
Richard Moore, 81
Marion Moses, 81
Peggy Shepard, 82
Cora Tucker (1940–), 83
Beverly Wright (1947–), 84
Concluding Note, 85

4 Documents, 87
    Laws, Treaties, Bills, and Executive Orders, 87
        The Civil Rights Act of 1964, 88
        The Fair Housing Act of 1968, 91
        National Environmental Policy Act of 1969, 92
        Statutory Civil Rights Requirements of the
            Environmental Protection Agency, 98
        Other Environmental Acts, 101
        The Environmental Equal Rights Act of 1993, 114
        Executive Orders, 119
        Arkansas State Law on Environmental Equity in Siting
            High-Impact Solid Waste Management Facilities, 127
        Basel Convention on the Control of Transboundary
            Movements of Hazardous Wastes and Their Disposal:
            Action in the U.S. Congress, 131
    Court Cases, 134
        *Margaret Bean et al. v. Southwestern Waste Management
            Corp. et al.*, 134
        *East Bibb Twiggs Neighborhood Association, et al. v. Macon-
            Bibb County Planning & Zoning Commission, et al.*, 137
        R.I.S.E., Inc., et al. v. Robert A. Kay, Jr., et al., 140

Washington, Mayor of Washington, D.C., et al. v. Davis et al., 142

Village of Arlington Heights et al. v. Metropolitan Housing Development Corp. et al., 144

Harrisburg Coalition Against Ruining the Environment v. Volpe, 146

El Pueblo Para el Aire y Agua Limpio v. Chemical Waste Management, 149

Recommendations and Policy Statements, 151
Recommendations to the Presidential Transition Team for the U.S. Environmental Protection Agency on Environmental Justice Issues, 151
Our Calls to Action, 155
Principles of Environmental Justice, 156
Comments to and About the EPA Environmental Equity Workgroup, 158
Model Environmental Justice Act, 162
Commission on Racial Justice, United Church of Christ, 171

5 Directory of Organizations, 173

6 Selected Print Resources, 207
Reports, 233
7 Selected Nonprint Resources, 235

8 Glossary, 243
Acronyms and Abbreviations, 259

Index, 263

About the Author, 273

Preface

Environmentalism and the civil rights movement have been two of the most important social movements in the United States over the past half century. At first glance, these two great movements would appear to have little or nothing in common with each other. And, in fact, some observers have argued that steps taken to improve the nation's environmental quality may actually have had a deleterious effect on the social and economic status of African Americans, Hispanic Americans, Asian Americans, and Native Americans. Studies have shown, for example, that laws passed to reduce environmental impacts overall have, in some cases, had an unfair impact on minority or poor communities.

Only within the last two decades has it become apparent that environmentalism and civil rights do have a great deal in common. That realization has come about primarily as the result of redefining the "environmental issues" to which Americans should turn their attention. More and more individuals and groups are pointing to the special environmental problems facing minority and low-income communities: hazardous waste sites; polluting industries; occupational hazards; and exposure to lead and other toxic metals, for example. The emphasis seems to be shifting from an almost exclusive concern with issues such as wilderness preservation and protection of endangered species to maintenance of livable environments for people of color and low-income communities.

Out of this view of environmental issues has grown a new movement—the environmental justice movement—that attempts to analyze patterns of disproportionate exposure to environmental hazards experienced by minority and low-income communities, to understand how such patterns have developed, and to develop programs by which disproportionate exposures can be remedied and prevented.

This book examines the nature and growth of the environmental justice movement. Chapter 1 provides a general introduction to the movement, explaining the forces that led to its birth, the philosophy that underlies much of its work today, and the tools through which it attempts to achieve its goals. Chapter 2 contains a chronology of the movement, a movement that, although still young, has recorded some important steps forward. Chapter 3 consists of biographical sketches of some important figures in the movement. Chapter 4 contains documents relating to the growth and work of the environmental justice movement, including relevant laws, court decisions, recommendations for action, and proposed legislation.

Chapter 5 provides a list of organizations active in the field. The list is representative only; one characteristic of the environmental justice is its emphasis on action by local and regional groups, which precludes the ability to provide an exhaustive list. Chapter 6 contains a selected bibliography of books and government reports relating to the environmental justice movement, while chapter 7 contains a similar list of nonprint resources. A glossary of terms commonly used in dealing with environmental justice issues is included, and an index is also provided.

The Road to Environmental Justice

Warren County, North Carolina: 15 September 1982

The protest tactics seemed clear enough. Most residents of Warren County knew how civil rights battles of the 1960s had been fought: with civil disobedience and nonviolent protest. The issue in 1982 was somewhat different, involving the construction of a waste disposal dump for toxic chemicals; but poor African Americans in Warren County knew their history well enough. The road to follow was not one of violent confrontation, but of nonviolent demonstration and protest.

The issue had arisen four years earlier when the Ward Transformer Company of Raleigh had illegally and surreptitiously disposed of 31,000 gallons of toxic polychlorinated biphenyls (PCBs) along 240 miles of roadways in 14 North Carolina counties. When the state of North Carolina uncovered the action, it was faced with the problem of digging up and relocating 40,000 cubic yards of contaminated soil. The state decided to bury those wastes in rural Shocco Township in Warren County.

The state's decision came under criticism almost immediately. Geological studies showed that soil under the proposed landfill was not impervious to leaching from the landfill, as required by U.S. Environmental Protection Agency (EPA) regulations, and the water table in the area was only about 7 feet below the landfill bottom, 43 feet shallower than required by the EPA. Since the majority of Warren County residents obtain their water from wells, these geological conditions raised serious questions about the state's decision. On 4 June 1979, however, the EPA gave North Carolina a waiver on both of these requirements and issued a permit for the landfill.

Residents of Warren County had had little experience in fighting political battles. As a group, they were poor (the county ranked 97th among North Carolina's 100 counties in per capita income) and African American (75 percent in Shocco Township and 64 percent in Warren County), and lacked in the resources, knowledge, and skills to fight a state political and bureaucratic system. Still, they used the skills they had. They made public appeals, called for help from national leaders of civil rights and environmental groups, and, in the end, prostrated their bodies in front of trucks carrying PCB-contaminated soil to their community. In all, 523 protestors were arrested.

In one sense, the Warren County protest failed. Over a six-week period, the 20-acre landfill was filled with 7,223 truckloads of wastes. The wastes are still in place, although leaching contamination into the underlying soil. In a larger sense, however, the protest was more successful than anyone could have imagined. For the first time in history, poor African Americans banded together—with the support of civil rights and environmental groups on a national level—to fight an environmental problem affecting a poor minority community (LaBalme 1987).

Environmental Inequities, Environmental Racism, and Environmental Justice

Most great social movements in history can be marked by a "defining moment," a single event so powerful and moving that later generations can look back and say, "It all started at. . . ." Rosa Parks' refusal to give up her seat on a Montgomery bus to a white man and move to the back of the bus is one example of such a defining moment in the civil rights movement. Historians may well look back at the protest by Warren County citizens and say, "That protest was the defining moment of the environmental justice movement."

Of course, such views of history are usually overly simplified and often incorrect. A nascent civil rights movement had been active long before Parks' heroic decision to confront racist laws in Montgomery, Alabama. And the seeds of a movement that would confront environmental inequities had been planted years before the Warren County protest brought the movement to public attention.

The issue facing residents of Warren County, North Carolina, in 1982 was and is not unique in the United States. The issue has been described as environmental inequity or environmental racism, terms that refer to the generally accepted evidence that environmental hazards are not distributed equally among various groups of people, either in the United States or throughout the world. Instead, communities of color and, to a lesser extent, poor people in general are exposed to hazardous and toxic wastes, dangerous working conditions, polluted air and water, and other environmental insults to a greater degree than are non-colored communities and people of higher economic status. Although the reasons for this pattern are disputed, its existence has been confirmed by multiple research studies.

The term *communities of color,* as described by those active in the environmental justice movement, refers to people whose skin is not white: African Americans, Hispanic Americans, Asian Americans, and Native Americans, for example. Although this definition does not correspond to that used by the U.S. Census Bureau, it is universally accepted within the environmental justice movement.

The terms used to talk about this phenomenon are important because they focus on and emphasize quite different aspects of the issue. *Environmental inequity* refers to a geographic reality, a pattern in which hazardous waste sites, polluting industries, nuclear waste dumps, and other environmental threats are more likely to be located within or adjacent to communities of color or poor communities. The existence of environmental inequities in the United States, within individual states, or in other localities can be determined—at least in principle—by research studies. Researchers count the number of environmentally hazardous facilities and determine whether they are likely to occur in such communities. Such studies have been conducted, and the evidence implies that environmental inequities do exist in the United States. A number of those studies, along with some that dispute the existence of environmental inequities, are discussed later in this chapter.

Environmental racism is a term that introduces a second concept into the discussion of environmental inequities. Environmental racism goes beyond acknowledging that such inequities exist and suggests a reason for their existence: racism. The phrase environmental racism was first used by Benjamin Chavis in 1982, then Executive Director of the National Association for the Advancement of Colored People, in connection with the protest at Warren County. He later explained precisely what he meant by the term during testimony before the U.S. House of Representatives Subcommittee on Civil and Constitutional Rights on 3 March 1993:

> Environmental racism is defined as racial discrimination in environmental policy making and the unequal enforcement of environmental laws and regulations. It is the deliberate targeting of people of color communities for toxic waste facilities and the official sanctioning of a life-threatening presence of poisons and pollutants in people of color communities. It is also manifested in the history of excluding people of color from the leadership of the environmental movement (U.S. Congress 1993).

A second term, *environmental discrimination*, is also used in referring to the unequal distribution of environmental insults, but may suggest that such inequalities affect communities other than those of color, such as those of low income. An important feature of both environmental racism and environmental discrimination is that they suggest that environmental inequities occur not because of chance, random events in history, but as the concrete and specific consequences of official public and corporate policies, conscious and deliberate or not.

Proving the existence of environmental racism or environmental discrimination is a quite different matter from proving the existence of environmental inequities. Even if one grants that such inequities exist, it does not necessarily follow that they are the result of some specific intent on the part of an individual or corporation. The debate as to whether environmental inequities occur as the result of concrete policies and practices designed to bring them about or as the incidental consequences of other business practices will be reviewed later in this chapter.

The terms *environmental justice* and *environmental equity* allude to yet another aspect of this issue. They refer to policies and practices by which existing environmental inequities can be corrected and prevented in the future. They focus on research programs that attempt to detect the existence of environmental racism and environmental discrimination; that uncover the underlying reasons that hold such practices in place; and that promote the enforcement of existing laws and regulations, the adoption of new rules and regulations, and the changes in philosophies and attitudes that are needed to eliminate environmental racism and environmental inequities from society.

Dr. Robert Bullard, one of the most prolific and articulate writers on the subject of environmental justice, has identified three broad categories into which the field of environmental justice can be subdivided. These are procedural equity, geographic equity, and social equity. Procedural equity refers to questions of fairness, "the extent to which governing rules and regulations, evaluation criteria, and enforcement are applied in a nondiscriminatory manner." Geographic equity refers to the location of environmental hazards with regard to communities of color and poor communities. Social equity concerns the way in which social factors, such as race, ethnicity, class, and political power, have an impact on and are reflected in environmental decision-making (Bullard 1994a).

The environmental inequities that exist in the United States include a variety of hazardous conditions. During the 1970s, much attention was focused on air and water pollution. Data from that period clearly show that the impact and costs of these environmental problems—although often not explicitly stated in reports—were not shared equally by all Americans. Two decades later, in the early stages of the environmental justice movement, research focused on municipal landfills, hazardous waste dumps, and emissions from chemical plants. As the movement has matured, however, it has become clear that many other forms of environmental inequities exist: in the siting of nuclear waste sites; in the nature and extent of occupational hazards to which workers are exposed; in the health hazards to which people are exposed in their daily diets; in the international distribution of hazardous and toxic wastes; and in the manufacture and sale of hazardous products.

Examples of Environmental Inequities

The term *racism* may cause one to focus on the inequalities that African Americans have historically experienced in the United

States. It is clear, however, that other groups within American society—including Native, Asian, and Hispanic Americans—have also been subjected to similar kinds of discrimination. The examples on the following pages are a few of the hundreds that could illustrate the environmental insults to which minority and low-income groups within American society have been exposed. The cases illustrate the disproportionate exposure of minorities to environmental hazards; however, they do not in and of themselves necessarily prove the existence of environmental racism.

West Dallas, Texas

West Dallas is a predominantly poor African American section of Dallas, Texas, with a 1987 population of 13,161, of whom 85 percent were African American. When the area was annexed by Dallas in 1954, a number of private homes were torn down as part of a "slum clearance" program to make way for large public housing projects. As a result of this program, the Dallas Housing Authority became the largest single landlord in West Dallas.

Immediately adjacent to the authority's 3,500-unit West Dallas housing projects is the 63-acre West Dallas RSR lead smelter, originally the Murph Metals secondary lead smelter. At its peak operation during the 1960s, the smelter released more than 269 tons of lead into the air annually. A significant fraction of the lead was blown across the homes and recreational areas of West Dallas.

In theory, such emissions came under the control of a strong lead emissions ordinance passed by the Dallas city council in 1968. According to Robert D. Bullard, however, "city officials systematically refused to enforce (the city's) lead emission standards" (Bullard 1994a). In support of this observation, studies found blood lead levels in West Dallas children to be 36 percent higher than those in children in control areas, a result attributed to the children's exposure to smelter emissions.

The period between the mid-1960s and the early 1980s was marked by the continuing release of lead particulates from the smelter, multiple studies showing elevated blood lead levels in West Dallas children, ongoing efforts by citizens of West Dallas to gain relief from their exposure to polluted air, and persistent stonewalling by local and national officials in the enforcement of existing regulations.

The issue was resolved in 1983 when the state of Texas and the city of Dallas sued RSR for violations of city, state, and federal lead emission standards. The case was settled out of court when RSR agreed to clean up soil in the West Dallas area that had been contaminated by lead emission from its smelter, to conduct a blood lead screening program for children and pregnant women in West Dallas, and to install antipollution equipment on its West Dallas plant. The final point of the agreement became moot four months later when the smelter was closed permanently.

The final word in this story may have been written in 1985 when RSR also agreed to a settlement in a civil case brought on behalf of 370 children in West Dallas. The company agreed to a program of payments that may eventually reach $45 million to those harmed by its smelter's emissions. Although the cash settlement is large it can never, as Bullard has pointed out, "repay the harm caused by lead poisoning of several generations of West Dallas children." On the other hand, it does send a message, he says, "that poor black communities are no longer willing to accept other people's pollution" (Bullard 1994a; Robinson 1994).

Churchrock, New Mexico

Churchrock Chapter is a governmental unit within the Navajo Nation, the largest Native American tribe. The Chapter is located east of Gallup, New Mexico, in an area characterized by extremely dry conditions with an average annual rainfall of about 7 inches. The major source of water in the region is the Rio Puerco, a stream that runs only intermittently, when fed by snow runoff or rainstorms.

Churchrock is the site of the longest continuous period of uranium mining in the Navajo Nation. That period began in 1954 with the discovery of uranium deposits along the north edge of the Rio Puerco Valley and continued until 1986. Mining rights were leased to uranium mining companies by the Navajo tribal government, but without the consent or participation of individual Navajo families living in the area. During most of the 30-year period during which mining continued, residents of the Churchrock Chapter had relatively little information about the environmental effects caused by uranium mining and milling in their area.

One of the most important of those effects was the removal of water from the Morrison Formation aquifer near the head of the Rio Puerco. Mining companies pumped water out of the aquifer at the rate of 5,000 gallons per minute in support of the construction and operation of underground mines for the removal of uranium.

That water was then returned to the Rio Puerco, converting the river into a continuously flowing stream.

This practice had two important effects on the Churchrock environment. First, it resulted in a significant loss of underground water from which many people drew water by means of wells. Second, it resulted in the release of water contaminated with radioactive wastes into the Rio Puerco, the region's major water source.

For years, the two companies mining uranium in the Churchrock region—the Kerr-McGee Nuclear Corporation and United Nuclear Corporation—argued that the Federal Water Pollution Control Act did not apply to their operations since the Rio Puerco was located on Native American lands and was not, therefore, a part of the United States. It was not until 1980 that the courts confirmed that the two companies were subject to the same Clean Water Act jurisdiction as were companies operating in other parts of the United States.

In 1983, the staff of the Southwest Research and Information Center in Albuquerque, New Mexico, examined Kerr-McGee and United Nuclear records to determine the extent to which the two companies were in compliance with federal Clean Water Act regulations. They found that Kerr-McGee was out of compliance 7 of the 35 months studied and United Nuclear was out of compliance 13 of 38 months in one location and 25 of 37 months in a second location. "Out of compliance" meant that mines were releasing waters that contained a higher concentration than permitted of dissolved uranium and/or radium-226, carried an excess of suspended solids, had a pH (a measure of the water's acidity or alkalinity) that was higher or lower than permitted, or failed to meet other water quality standards.

As unsatisfactory as these records were, they were dwarfed by a serious accident: the failure of a United Nuclear uranium mill tailings dam on 16 July 1979. More than 94 million gallons of contaminated liquids with a pH of 1 (roughly comparable to that of battery acid) were released into the Rio Puerco. The spill posed a serious health threat not only to Native Americans living in the area, but also to the livestock on which many Native Americans depended for their livelihood. In 1985, United Nuclear settled a lawsuit brought to it as a result of this accident by agreeing to pay $550,000 to 240 plaintiffs, an award of about $2,000 per plaintiff.

Still, it was not the 1979 spill that has been the most serious problem for residents of the Churchrock area. William Paul Robinson, Research Director of the Southwest Research and

Information Center, has written that the spill "does not appear to have had as devastating an effect on the Rio Puerco as the decades of mine dewatering (the removal of water that seeps into a mine as it is dug), which preceded the spill. Studies of human and livestock health effects after the spill indicated that the same pollutants found in high concentrations in mine and tailings water had shown up in abnormally high levels in the muscles and organs of cattle, sheep, and goats that grazed along the Rio Puerco downstream of the mines and mills" (Robinson 1992).

As is often the case in environmental justice cases, the long-term effects of uranium mining in the Churchrock area may be more positive than one might have imagined. Residents have formed the Puerco Valley Navajo Clean Water Association to learn more about the dangers of polluted waters and to search for new and safer water supplies for their community. As Robinson has pointed out, the Association's tenth anniversary celebration on 16 July 1989 "focused on the need for water in the communities, not the horror of the spill itself" (Robinson 1992).

"Cancer Alley"

In some instances, a pattern of environmental inequity may extend over broad regions, affecting many communities at once. Such is the case with an 85-mile stretch of land along the Mississippi River between Baton Rouge and New Orleans, Louisiana. Louisiana has long publicized itself as a "sportsman's paradise" and depended on agriculture and fishing as its main sources of income. At the end of World War II, however, the state's political economy began to change. Attracted by the potential for easy transportation and waste disposal as well as the availability of cheap labor, chemical and petroleum companies moved into the region between the state capitol and its largest city. Over the decades, these companies dramatically changed the character of the region.

Today, more than 125 companies producing fertilizers, gasoline, paints, plastics, and other chemical products line the banks of the Mississippi. Approximately one-quarter of the nation's petrochemicals are produced in the region. In addition, the plants release huge amounts of toxic wastes to the air, water, and surrounding countryside. By one estimate, more than 2 billion pounds of toxic chemicals were emitted in the two-year period between 1987 and 1989 (Louisiana Advisory Committee 1993). It is for these reasons that the region has been given the name

"Cancer Alley" by residents of the area. The *Washington Post* has also referred to the region as a "massive human experiment" and "a national sacrifice zone" ("Jobs and Illness" 1989).

The development of the new industrial complex along Cancer Alley, while undoubtedly beneficial to the state economy, has caused severe dislocations among communities along the river. In a 1993 study of Cancer Alley, the Louisiana Advisory Committee to the U.S. Civil Rights Commission selected eight communities to study in detail: Revilletown, Sunrise, Morrisonville, Alsen, Wallace, Forest Grove, Center Springs, and Willow Springs. The committee found that the interaction between corporations and communities differed from town to town and the ultimate resolution of disputes between the two entities also differed from case to case.

In some instances, residents were able to prevent a new plant from moving into an area. In such cases, the community was able to survive, although it was not necessarily spared a significant increase in exposure to air- and water-borne pollutants released by other nearby plants. In other cases, towns acceded to offers from companies to sell their homes and land in order to make room for new plants. Two of the communities studied, Revilletown and Sunrise, were dismantled completely as a result of buyout programs and a third, Morrisonville, was relocated to make room for a plant.

Despite the accommodations made by individual communities, the committee came to the conclusion that the 85-mile stretch of Cancer Alley had, indeed, earned its name. It reported to the U.S. Civil Rights Commission that

> . . . black communities in the corridor between Baton Rouge and New Orleans are disproportionately impacted by the present State and local government systems for permitting and expansion of hazardous waste and chemical facilities. These communities are most often located in rural and unincorporated areas, and residents are of low socioeconomic status with limited political influence.

The committee further concluded that a major reason for the existence of Cancer Alley was that

> . . . state and local governments have failed to establish regulations or safeguards to ensure such communities

are reasonably protected from a high concentration of hazardous waste and industrial facilities and risks associated with living in and around such facilities (Louisiana Advisory Committee 1993).

Pesticide Exposure

For the past 50 years, U.S. farmers have relied heavily on the use of synthetic chemicals to protect their crops from destruction by insects, rodents, microorganisms, and other predators. Names such as DDT, aldrin, malathion, parathion, chlordane, heptachlor, toxaphene, and Sevin are now familiar to many Americans whether they are engaged in agriculture or not. These chemicals have been popular with farmers because the chemicals kill pests effectively at relatively modest costs.

As the use of synthetic chemicals spread through agriculture, however, some important disadvantages began to occur. Among other problems, some of the most popular pesticides proved to be toxic not only to pests, but also to harmless or beneficial animals and to humans themselves. As a result, government agencies began to monitor the use of pesticides much more carefully, actually banning some of the most dangerous pesticides. Over time, many Americans began to feel that the most serious environmental problems posed by synthetic pesticides were being brought under control.

However, pesticides continued to pose dangers to farmworkers. Many of the regulations passed to protect consumers from pesticides made no mention of the dangers posed to men and women working in the fields. For example, the EPA banned the use of many chlorinated hydrocarbon pesticides, such as DDT, aldrin, dieldrin, heptachlor, and chlordane in 1972. This class of pesticides tends to be persistent, meaning that they remain on fruits and vegetables long enough that they may be present when consumed by the general public. Farmers switched instead to pesticides known as organophosphates (malathion, parathion, and methyl parathion, for example) and carbamates (Sevin, Zectran, and Temik, for example). These compounds are more toxic, but less persistent. They pose less threat to consumers (because they have degraded by the time fruits and vegetables reach the marketplace), but greater threat to farmworkers (Miller 1985).

The problem of pesticide exposure is an important issue in the environmental justice movement because a large proportion of farmworkers are Hispanic Americans, especially Mexican Americans. Many are illegal aliens who work for less-than-minimum wages, often under difficult and illegal working conditions. They are seldom in a position to complain about working conditions since objections are likely to be met with dismissal from their jobs.

It is true that laws have been passed to protect farmworkers. In general, however, those laws have been weak, poorly enforced, and, therefore, largely ineffective. For example, it is illegal to use pesticides on a field in such a way that workers will be sprayed directly. But this regulation does not protect workers on an adjacent field from being exposed to pesticides blown onto them. Also, state officials are allowed to use otherwise unlicensed pesticides in case of an emergency, such as the unexpected appearance of a pest. In such a case, the implication is that it is acceptable to expose farmworkers to a chemical hazard to save crops.

Some regulations exist governing the period of time after spraying during which workers are not allowed to reenter fields. But given the paucity of data on safe intervals for most pesticides, such regulations may not be effective. Overall, one observer has concluded that "the reality [of existing pesticide legislation] is that the extant legislation leaves farm workers virtually unprotected against pesticide hazards" (Perfecto 1992; also see Wasserstrom and Wiles 1985; Moses 1989; and Moses 1993).

Consumption of Toxic Fish from the Detroit River

Environmental inequities can exist in many forms. One that is probably less apparent than the cases described previously is in the dietary patterns of minorities who rely on fish taken from the Detroit River. An analysis of this issue has been carried out by Patrick C. West, associate professor in the School of Natural Resources at the University of Michigan, and his colleagues. The study was motivated by the fact that state and federal guidelines currently specify the amount of toxic contaminants that may be released to surface waters. The regulated level of contaminants is determined, in turn, by assuming that the average person will consume a certain amount of fish per day. For example, the State of Michigan's Rule 1057, which regulates the discharge of toxic chemicals into the state's surface water, assumes that the average level of fish consumption in Michigan is 6.5 grams per person per day. For example, under those circumstances, it might be permissible for there to be no more than 0.1 milligram of a toxic substance in the fish without endangering the health of the consumer.

The flaw in this logic, however, is that some residents of Michigan are likely to consume more than 6.5 grams of fish each day on a regular basis. A person who regularly eats 10 or 20 grams of fish, for example, would then be exposed to two or three times the toxic substance as that assumed by the regulatory agency. Instead of the consumer being protected by the federal or state regulation, he or she might be lulled into a false sense of safety.

West and his colleagues attempted to find out the extent to which this situation might exist in the State of Michigan and, if it did, the extent to which certain individuals might be exposed to unexpected environmental risks from eating fish. West's team made two interesting discoveries. First, they found that among those who fish in the Detroit River, 21.7 percent of whites in the sample were more likely to fish for both recreational purposes and to obtain food for meals, while 58 percent of nonwhites fished for both recreational purposes and to obtain food.

In addition, the West research team found that members of all minority groups studied (African Americans, Native Americans, and other minorities) consumed about five times the amount of fish per day as assumed by Michigan's Rule 1057. The team concluded that the Rule was, therefore, inadequate to protect those who fish primarily for food rather than for recreation. They ended their report of this research with the observation that "To rely on a policy of fish consumption advisories creates more of a hardship for minorities because a needed protein source is at stake, not just a chance to catch that big one that didn't get away" (West 1992; West et al. 1992).

Environmental Inequities
on an International Level

Gammalin 20 seemed to be a blessing from the heavens for the fishermen in the village of Achedemade Bator along the shores of Lake Volta in Ghana. Although the chemical had been imported for use as a pesticide, the fishermen found another use for the product. By dumping Gammalin 20 into the lake, they easily killed as many fish as they could possibly want. All they had to do was to go out in their boats and pick up the dead fish floating on the surface of the water. For a village that depended on the lake for both its drinking water and its main source of protein (the fish), Gammalin 20 was truly a wonderful discovery.

What the fishermen of Achedemade Bator did not know was that Gammalin 20, also known as lindane, is a highly toxic relative of DDT. It may cause dizziness, headache, nausea, vomiting, diarrhea, tremors, weakness, convulsions, and circulatory disorders. Long-term exposure to the chemical can cause liver damage and, according to *The Merck Index*, it "may reasonably be anticipated to be a carcinogen (The *Merck Index* is a highly regarded and widely used encyclopedia of chemicals, drugs, and biological substances.)." Lindane is so highly toxic, in fact, that it has been banned for most uses in the United States and other developed countries of the world. Its export to less developed countries, such as Ghana, however, was still legal in the late 1970s when the fishermen of Achedemade Bator were using the pesticide as a simple form of catching fish.

It did not take long for the effects of Gammalin 20 to show up in the lake. The fish population began to drop off at the rate of 10 to 20 percent per year, and some villagers began to experience the symptoms of lindane poisoning, although they made no connection between their illnesses and the use of Gammalin 20. Finally, a Ghanian organization known as Association of People for Practical Life Education came to the rescue of Achedemade Bator. The group explained to villagers about the dangers of Gammalin 20 and encouraged them to return to traditional forms of fishing with nets and traps. Eventually, the fish population returned to normal and the strange illnesses affecting residents of Achedemade Bator disappeared (Norris 1982).

The lesson learned at Achedemade Bator is one that has been repeated over and over again in the last few decades. Chemicals produced for use as pesticides, medications, fabrics, dyes, and building materials are found too hazardous for use by the general public in the United States or another developed country and are banned completely or for all but very limited uses. Those chemicals are still allowed to be exported to less developed nations, however, where health and safety standards are less rigorous than they are in the developed world. Similarly, hazardous or radioactive wastes considered too dangerous to be stored in the United States or another developed nation are shipped to other parts of the world where they result in unknown numbers of illness and death among native populations.

These forms of environmental inequities are as much forms of environmental racism as are the problems faced by residents of

Warren County, North Carolina; farms in Churchrock, New Mexico and elsewhere in southwestern United States; or the lakes and rivers of Michigan. (See also Bullard 1994b.)

Environmental Justice: Civil Rights + Environmentalism

The movement for environmental justice represents the confluence of two older movements in the United States—the environmentalist and civil rights movements. The origins of the environmentalist movement in this country can be traced to the late nineteenth century. Before then, concern about the environment was virtually nonexistent in the United States. Instead, the dominant philosophy was that of the Biblical injunction to "go forth and conquer the Earth." Early pioneers sweeping westward from the original 13 colonies encountered unimagined riches of land, minerals, timber, and wildlife. They harvested and used those resources without consideration for preserving or conserving them for some future use. Indeed, official governmental policy encouraged this rape of the land and its resources. The Preemption Act of 1841, for example, allowed a settler and his family to buy homestead land of up to 480 acres for 50 cents an acre. In many cases, lumber companies hired gangs of men to claim the land and hand it over to the companies. A similar practice was followed by cattle barons, who hired cowboys to claim homesteading lands under the Desert Land Act of 1877 and then use the lands for grazing their cattle (Petulla 1977).

It was only with the end of the westward migration, as pioneers reached the Pacific Coast, that some individuals realized that the continent's resources were not unlimited. This realization spurred the first environmental movement in the United States.

That movement actually consisted of two themes—conservationist and preservationist. The term *conservation* refers to the practice of harvesting natural resources in a controlled manner so that they will be available for future generations. The current U.S. government policies of "multiple use, sustained yield" reflect this philosophy. Conservationists do not argue that nature is sacred and inviolable, but that it provides humans with many valuable resources that we must learn to use wisely. One of the most articulate spokespersons for the early conservationist movement was Gifford Pinchot, Chief of the Division of Forestry (later the National Forest Service) in the Department of Agriculture (Wild 1979).

In contrast to the views of the conservationists was the philosophy of preservation. Preservationists argue that there is a beauty and value in nature that has nothing to do with the commercial value for humans. Humans have the obligation to protect vast portions of the natural world for no other reason than inherent value. According to preservationists, large segments of the world's forests, deserts, grasslands, and other natural resources should be set aside forever, therefore, for their own protection and for their enjoyment by humans. The philosophy of preservation owes much to the early writings of Henry David Thoreau. Thoreau once wrote that "there is a subtle magnetism in Nature, which, if we unconsciously yield to it, will direct us aright" (Thoreau 1962). Perhaps the most famous preservationist of all was John Muir, one of the founders of the Sierra Club and the man largely responsible for convincing President Theodore Roosevelt of the need for the U.S. National Park system.

The traditional environmental movement, then, was one that strongly emphasized the connection between humans and nature. Hiking, backpacking, mountain climbing, bird-watching, photography, canoeing, and many other forms of recreation were the preferred activities of those who called themselves "environmentalists" until the mid-twentieth century. The environmental groups that sprung up within this movement—groups such as the Sierra Club, the Wilderness Society, and the National Parks and Conservation Association—strongly reflected that emphasis. Their goal was to protect the natural environment for a variety of human uses, whether for pure enjoyment or limited commercial exploitation.

The Modern Environmental Movement

The 1960s saw the rise of a quite different kind of environmental movement. This movement was inspired by a dawning recognition of the havoc humans were wreaking on the natural environment. Some people credit Rachel Carson's book *The Silent Spring* as the defining moment in the modern environmental movement. Carson showed how the indiscriminate use of pesticides killed insects and the birds who fed on them (thus causing "the silent spring"). She called attention to the fact that many of the wonders of modern science and technology that made life healthier, safer, and more pleasant for humans were, at the same time, causing extensive damage to the physical and biological environment.

Out of this dawning realization grew a host of new environmental groups, groups whose focus was not primarily the enjoyment of nature, but the battle focusing on issues such as air and water pollution, hazardous waste disposal, and land use issues. Examples of such groups are the Acid Rain Foundation, the Citizens' Clearinghouse for Hazardous Wastes, the Environmental Action Foundation, and the Friends of the Earth.

Groups that make up the traditional and modern environmental movements are sometimes said to constitute mainstream environmentalism. Early activists in the environmental justice movement have often argued that they have little in common with mainstream environmentalism. One reason for this position is that mainstream environmentalist groups tend to be largely white and middle- or upper-class. One study of the membership of mainstream environmental groups in the 1940s, for example, found that 96 percent of the 1,468 respondents classified themselves as Caucasian/European. Almost half (48 percent) had a total family income of more than $10,000, and 15 percent had a total family income of more than $25,000. For comparison, only 10 percent of American families made $13,000 or more at the time of the survey (as cited in Smith 1974).

In addition, the evidence suggests that most mainstream environmentalist groups have traditionally had little interest in issues faced by poor, minority, urban people. When members of the Sierra Club were asked in the 1970s, for example, whether the club should concern itself with the conservation problems of such special groups as the urban poor and ethnic minorities, about 40 percent said "no," and about 15 percent said "yes" strongly (as cited in Smith 1974). When a similar poll was conducted more than a decade later, a proposal to increase involvement in environmental issues faced by the urban poor and communities of color was defeated by a vote of about three to one (Freudenberg 1984).

Under the circumstances, it is hardly surprising that at least some observers have had harsh words for mainstream environmentalism. For example, in his book *Environmental Quality and Social Justice in Urban America*, James Noel Smith has argued that mainstream environmentalism is "a deliberate attempt by a bigoted and selfish white middle-class society to perpetuate its own values and protect its own lifestyle at the expense of the poor and the underprivileged" (Smith 1974). A similar view was expressed even earlier by Richard Hatcher, then mayor of the city of Gary, Indiana. "The nation's concern with the environment," Hatcher said, "has done what George Wallace had been unable to do: distract the nation from the human problems of black and brown Americans" ("The Rise of Anti-Ecology" 1970).

In fact, the accomplishments of mainstream environmentalism have sometimes exacerbated the environmental problems of low-income people and people of color. As an example, new regulations on the use of pesticides in agriculture, rules enacted under the pressure of mainstream environmental groups, have in some cases increased the risk to farmworkers. Agricultural business people have replaced pesticides of lower toxicity and longer persistence (thereby providing greater protection to consumers) with pesticides of greater toxicity and shorter persistence (thus increasing the risk to farmworkers) (Perfecto 1992).

In any case, many of those active in the environmental justice movement retain doubts about the ability of mainstream environmentalist groups to understand or respond to the issues about which people of color are most concerned. In assessing the current relationship between mainstream environmentalism and the environmental justice movement, Robert Bullard has written that

> The mainstream environmental movement has proven that it can help enhance the quality of life in this country . . . Yet, few of these groups have actively involved themselves in environmental conflicts involving communities of color. Because of this, it's unlikely that we will see a mass influx of people of color into the national environmental groups any time soon. A continuing growth in their own grassroots organizations is more likely (Bullard 1993).

The Civil Rights Movement

For some historians, the modern civil rights movement in the United States can be traced to 1 December 1955 when African American seamstress Rosa Parks refused to give up her seat in the white section of a Montgomery, Alabama, bus to a white man. The event precipitated a year-long boycott of Montgomery buses by African Americans in the city and culminated in a decision by the U.S. Supreme Court that segregation on public bus systems was illegal.

The next decade was marked by the battle by African Americans and their allies to eliminate the whole fabric of segregation that had been constructed in the South over the preceding century. Important events included the Alabama Freedom Riders of 1961, civil rights demonstrations in Birmingham, Alabama, and the march on Washington, D.C., in 1963, and the march on Selma, Alabama, in 1965. Although protests were strongly influenced by the philosophy of nonviolence, they were often met with extreme violence, resulting in injuries to and the deaths of hundreds of protestors.

The civil rights movement's major goals were slowly accomplished, however, with the passage of the Voting Rights Act of 1957, the Civil Rights Act of 1964, the Fair Housing Act of 1968, and similar legislation. Courts also confirmed that segregation of any kind in American society was unconstitutional and set forth rigorous programs for the elimination of bias in education, housing, employment, and other fields. Although the centuries-old evils of segregation were hardly eliminated, the nation's highest legislative and judicial bodies had made clear that official forms of apartheid in the United States were illegal (Grossman 1993; Morris 1984).

The modern environmental justice movement is, therefore, the child of two great social movements in American history: the environmental and civil rights movements. However, the environmental justice movement probably owes more to one parent, the civil rights movement, than to the other, the environmental movement. Although the issues with which environmental justice deals are environmental issues—air and water pollution, hazardous waste sites, and the like—the political philosophy that underlies action is a reflection of the civil rights philosophy. In addition, most of the groups currently acting on environmental justice issues began their existence as civil rights groups and not environmental organizations. As Robert Bullard has said,

> The push for environmental equity is an extension of
> the civil rights movement, a movement in which di-
> rect confrontation and the politics of protest have been
> real weapons (Bullard 1994a).

Environmental Justice Comes of Age

Walter E. Fauntroy, representative from the District of Columbia to the U.S. House of Representatives, participated in and was arrested for his part in the 1982 Warren County protest described previously. Following his release from jail, Fauntroy asked the U.S. General Accounting Office (GAO) to conduct a study on the relationship between pollution and minority communities. The report found that three of the four largest landfills located in the Southeast were located in predominantly poor and African American communities (GAO 1983).

The GAO study was rather modest in its scope, but it was important for another reason. It brought to the attention of the general public, probably for the first time, a possible association between race and income on the one hand and exposure to environmental hazards on the other. In addition, it highlighted the serious lack of fundamental research on this issue.

The Commission on Racial Justice of the United Church of Christ

In response to the gap in fundamental research, the Commission on Racial Justice of the United Church of Christ decided to initiate an even more extensive study of the environmental problems faced by minority and poor communities. The Commission analyzed the location of hazardous waste sites in all ZIP (postal) codes and counties in the United States. The results of the study were released in 1987 and were immediately hailed by experts in both environmental and civil rights communities. The *Atlanta Constitution* observed, for example, that the study "put an end . . . to speculation that white America has been dumping its garbage in black America's backyard." The study "puts an extra burden of responsibility on public health and environmental inspectors," the *Constitution* went on to say, "to spot potential problems before they become health hazards. Where to look? The Commission report points the way" ("Dumping on Black America" 1987).

What is it that the Commission on Racial Justice had actually found? Among the most important points in the Commission report were the following:

1. People of color are twice as likely to live in a community with a commercial hazardous waste site than are whites. They are three times as likely to live in a community with more than one hazardous waste site.
2. About 60 percent of African Americans live in a community with an abandoned hazardous waste site.

3. The average annual income of people living in communities with a hazardous waste site is significantly less than those living in communities without such a site.
4. In spite of the preceding point, race, more than income, is a more reliable factor in predicting the presence of a hazardous waste site.
5. Three of the five largest hazardous waste sites, accounting for about 40 percent of the nation's total landfill capacity, are located in communities that are predominantly African American or Latino.
6. Fifteen million African Americans and eight million Hispanic Americans live in communities with one or more hazardous waste sites.
7. Of the six cities with the largest number of hazardous waste sites, African Americans make up a far greater fraction of residents than do whites. The numbers for those cities are as follows: Memphis (43.3 percent African Americans; 173 sites); St. Louis (27.5 percent; 160 sites); Houston (23.6 percent; 152 sites); Cleveland (23.7 percent; 106 sites); Chicago (37.2 percent; 103 sites); and Atlanta (46.1 percent; 94 sites). For comparison, African Americans make up 11.7 percent of the general U.S. population (Commission for Racial Justice 1987).

The Commission concluded its report with a number of recommendations (see chapter 4).

As dramatic as was the Commission on Racial Justice report, it was certainly not the first serious study of the relationship among environmental hazards, race, and income. For example, an early study by David Harrison found that the costs of air pollution were more likely to be borne by low-income groups than by middle- or high-income groups (Harrison 1975).

An important study was that completed by Robert D. Bullard and reported in 1983. Bullard found that the pattern of siting for municipal landfills, incinerators, waste transfer stations, and other forms of waste disposal facilities in Houston, Texas, was similar to that later to be reported by the Commission on Racial Justice. That is, all five of the city-owned landfills are located in predominantly African American neighborhoods; six of eight garbage incinerators are located in such neighborhoods, one in a predominantly Hispanic American neighborhood and one in a predominantly white area (Bullard 1983). Over the past decade and a half, a number of studies similar to those of Bullard's have been conducted in specific cities, counties, and states (see, for example, Brueggemann 1993; Holtsman 1992; and Roberts 1992).

The Michigan Conference

As the decade of the 1990s dawned, environmental justice had become a topic of considerable interest across a wide spectrum of citizens from all walks of life. In 1990, two academicians introduced to the issues of environmental justice relatively recently— Bunyan Bryant and Paul Mohai, at the University of Michigan School of Natural Resources—convened a conference of scholars and activists in Ann Arbor to present their latest research, discuss their ideas, and consider possible solutions to the problems of environmental racism and environmental inequities. A number of papers presented at the Michigan Conference were later reprinted in *Race and the Incidence of Environmental Hazards*, edited by Bryant and Mohai.

The Michigan Conference was important because of other events that it initiated. First, participants at the conference drafted a letter to Louis W. Sullivan, Secretary of the U.S. Department of Health and Human Services, William K. Reilly, Administrator of the EPA, and Michael R. Deland, Chair of the Council on Environmental Quality. The letter asked for an opportunity to meet with government officials to discuss a number of issues relating to environmental justice. Among these issues were the request for more research on the nature of environmental inequities in the United States, the development of projects targeted at providing aid to low-income and minority populations, the appointment of special assistants for environmental equity in government agencies, and the development of a policy statement on environmental equity. Representatives of the Michigan Conference met with Reilly and other government officials on 13 September 1990 in what appeared to be a highly productive meeting.

Governmental Recognition

Although the administration of President George Bush took no concrete action to deal with the issues of environmental justice, an important seed had been planted. That seed came to fruition with the election of Bill Clinton as President in 1992. Clinton appointed Benjamin Chavis and Robert Bullard, both activists in the field of environmental justice, to serve on his Transition Team in the area of Natural Resources (which included the EPA and Departments of the Interior, Energy, and Agriculture). Then, on 11 February 1994, President Clinton issued Executive Order 12898, establishing an Office of Environmental Justice within the EPA and a National Environmental Justice Advisory Council (NEJAC).

Clinton's action is arguably the most important step to occur in the environmental justice movement in its short history. The creation of the Office of Environmental Justice and the NEJAC provided the mechanism and the financial support that makes it possible for those concerned about environmental justice issues to meet at regular intervals, design programs of research, present their concerns directly to the federal administration, hold public meetings at which groups and individuals can testify, and comment on federal programs that have a relevance to environmental inequities.

The NEJAC consists of about two dozen members selected from community-based groups, business and industry, academia, tribal governments, nongovernmental organizations, and environmental groups. NEJAC is organized into four subcommittees—Waste and Facility Siting, Enforcement, Health and Research, and Public Participation and Accountability.

First National People of Color Environmental Leadership Summit

Another important event in the organization of a national environmental justice movement took place in October of 1991. Under the auspices of the Commission on Racial Justice of the United Church of Christ, the First National People of Color Environmental Leadership Summit was convened in Washington, D.C. More than 500 participants from a wide variety of ethnic, racial, cultural, and economic groups came together to talk about their mutual concerns about the environment and environmental inequities. More than 50 presentations, workshops, panels, caucuses, and other sessions were held during the three-day conference. At the conference's conclusion, the participants adopted a 17-part statement, "Principles of Environmental Justice," outlining their combined view on the state of environmental inequities in the United States. The principles have become a guideline for much of the thought and work that now takes place in environmental justice groups throughout the nation. (See chapter 4 for a statement of the principles.)

Environmental Justice as a Social Movement

The environmental justice movement in the United States in the 1990s stands out among other social movements. Although it has begun to take on some of the trappings familiar to the movements from which it has sprung, namely a centralized organization located in large governmental and nongovernmental bodies, it still remains to a large degree a highly decentralized movement. While academicians and activities were coming together at the Michigan Conference and in the development of the EPA Office of Environmental Justice, dozens of small, local groups were being organized to deal with the very real issues faced by individual neighborhoods and communities. (For a superb discussion of all facets of the environmental justice movement, see Lazarus 1993.)

The 1990s saw the formation, for example, of groups such as The Coalition for Environmental Consciousness in Ridgeville, Alabama; California Indians for Cultural and Environmental Protection in Santa Ysabel, California; the Citizens League Opposed to Unwanted Toxins in Tifton, Georgia; Ke Kua'aina Hanauna Hou in Kaunakakai, Hawaii; the Flint-Genessee United for Action, Justice, and Environmental Safety in Flint, Michigan; Concerned Citizens of Sunland Park in Sunland Park, New Mexico; Fort Greene Community Action Network in Brooklyn, New York; Eufaula Street Landfill Committee in Fayetteville, North Carolina; Environmental Services Office of the Cherokee Nation in Tahlequah, Oklahoma; People Organized in Defense of Earth and its Resources in Austin, Texas; and the Environmental Center for New Canadians in Toronto, Ontario.

Representatives of such groups often stay in contact with each other and draw on each other's resources through tribal, regional, cultural, or other affiliations, such as the Southwest Network for Environmental and Economic Justice, the Asian Pacific Environment Network, the Gulf Coast Tenants Organization, or the National Congress of American Indians. Yet, in many cases, the bulk of their actual work occurs within a neighborhood, a small town, tribe or reservation, or a county.

Local groups also stay in contact with each other through a number of journals, magazines, newsletters, and other publications that have grown up over the past two decades. These include *Race, Poverty, and the Environment* (published by the Earth Island Institute), *Everyone's Backyard* (Citizens' Clearinghouse for Hazardous Wastes), *RACHEL's Hazardous Waste News* (Environmental Research Foundation), *Toxic Times* (National Toxics Campaign), *The Egg: A Journal of Ecojustice* (Network Center for Religion, Ethics, and Social Policy), *Voces Unidas* (Southwest Organizing Project), and *Panna Outlook* (Pesticide Action Network) (Taylor 1993).

An interesting feature of the modern environmental justice movement is the critical role played by women. Mothers, housewives, secretaries, and rural women have long been at the forefront of protests, community action, leafleting, public testimony, and other actions on issues concerning environmental inequities and environmental racism. A number of writers have observed that many issues in the field of environmental justice strike close to the individual home and thus arouse more interest among housewives and mothers than would issues from mainstream environmentalism, such as the protection of endangered species or the development of national parks.

Celene Krauss, Assistant Professor of Sociology and Coordinator of Women's Studies at Kean College of New Jersey, for example, has drawn on interviews with women active in the environmental justice movement to place their participation in a broader context of feminist political philosophy. "Female blue-collar activists," she concludes, "often share a loosely defined ideology of environmental justice and a critique of dominant social institutions and mainstream environmental organizations, which they believe do not address the broader issues of inequality underlying environmental hazards. At the same time, these activities exhibit significant diversity in their conceptualization of toxic waste issues, reflecting different experiences of class, race, and ethnicity" (Krauss 1994).

Data and Statistics

Evidence that environmental hazards are distributed disproportionately in the United States among communities of various races and incomes is now widely, although not universally, accepted. Perhaps the best single summary of the evidence on this issue can be found in Benjamin A. Goldman's *Not Just Prosperity: Achieving Sustainability with Environmental Justice.* Goldman's report was prepared for use at the National Wildlife Federation's Corporate Conservation Council conference on "Synergy '94: The Community Responsibilities of Sustainable Development," held in February 1994 (Goldman 1993). The following summary draws heavily on the information presented in Goldman's report.

One of the earliest studies dealing with the disproportionate distribution of environmental inequities was that of Hoffman et al. in 1967. Those researchers found that nonwhites in the sample they studied had accumulated a larger concentration of pesticides in their bodies than had a comparable sample of whites (Hoffman et al. 1967). Since the Hoffman research, more than 100 other studies on environmental inequities have been completed. Those studies can be categorized on the basis of various criteria, such as the field from which researchers have come (economics, sociology, health sciences, environmental science, and urban planning, for example), the organizational support that made possible the research (academic institution, governmental agency, or industry, for example), the temporal perspective of the research (prospective versus retrospective), the time period covered by the study, the research method used, the geographical region covered (nation, region, state, county, or ZIP code area, for example), and the environmental concern of the study (hazardous or nonhazardous wastes, air or water pollution, blood levels, or diet, for example). Table 1 summarizes some of the general characteristics of the studies reviewed by Goldman (1993).

Race, Economic Status, and Environmental Inequities

Probably of greatest interest are the last two of these classification systems, geographic focus and environmental concern. Goldman's summaries for these two variables are reproduced in Tables 2 and 3. The most important single conclusion to be drawn from these summaries is the environmental inequities that exist on the basis of race and income. Only one of the 64 studies examined by Goldman did not find an environmental disparity based either on income or on race. That study, conducted under

Table 1 Some General Characteristics of Studies Reviewed by Goldman

1. Decade Conducted:
1960s: 1
1970s: 18
1980s: 19
1990s: 26

2. Background of Researchers:
Epidemiology: 13
Environmental scientists: 5
Economics: 12
Journalists: 5
Government agencies: 8
Nonprofit organizations: 5
Geographers: 6
Sociologists: 3
Others (including legal scholars, urban planners, industrial corporations, political scientists, and psychologists): 7

3. Geographic Scope:
National: 28
Selected states or regions: 8
Selected cities or counties: 24
All urban areas: 4

4. Environmental Concern:
Hazardous waste: 16
Solid waste: 8
Air pollution: 11
Pesticide exposure: 5
Regulatory costs and benefits: 9
Toxic fish consumptions: 5
Blood lead levels: 2
Toxic and radioactive releases: 9
Fatal hazardous exposure: 1
More than one concern: 7

Adapted from Goldman 1993.

Table 2 Summary of Findings of 64 Empirical Studies of Environmental Disparities by Income and Race in the United States by Environmental Concern*

Environmental Concern	Racial or Income Disparities/ Number of Studies	Racial Disparities/ Number of Tests	Income Disparities/ Number of Tests	Race More Important than Income/ Number of Tests
Hazardous waste	15/16 (94%)	14/18 (78%)	13/17 (77%)	7/8 (88%)
Air pollution	11/11 (100%)	9/10 (90%)	9/10 (90%)	4/7 (57%)
Regulatory costs and benefits	9/9 (100%)	4/4 (100%)	9/10 (90%)	3/3 (100%)
Toxic and radioactive releases	9/9 (100%)	8/9 (89%)	6/8 (75%)	4/6 (67%)
Occupational health	8/8 (100%)	8/8 (100%)		
Solid waste	8/8 (100%)	11/15 (73%)	8/13 (62%)	4/6 (67%)
Pesticide exposure	5/5 (100%)	4/4 (100%)	2/2 (100%)	
Toxic fish consumption	5/5 (100%)	5/5 (100%)	0/2 (0%)	1/1 (100%)

Table continued on next page

Table 2 Summary of Findings of 64 Empirical Studies of Environmental Disparities by Income and Race in the United States by Environmental Concern* (continued)

Environmental Concern	Racial or Income Disparities/ Number of Studies	Racial Disparities/ Number of Tests	Income Disparities/ Number of Tests	Race More Important than Income/ Number of Tests
Blood lead levels	2/2 (100%)	1/1 (100%)	1/1 (100%)	1/1 (100%)
Fatal hazardous exposures	1/1 (100%)	1/1 (100%)		
TOTAL	63/64 (98%)	59/68 (87%)	43/58 (74%)	22/30 (73%)

Reprinted by permission of the Corporate Consumer Council of the National Wildlife Federation from Goldman 1993.
* Some of the studies and tests involved more than one environmental concern, so the numbers in each column do not necessarily add up to totals in the bottom row. Blanks indicate tests were not performed.

Table 3 Summary of Findings of 64 Empirical Studies of Environmental Disparities by Income and Race in the United States by Geographic Unit of Analysis*

Geographic Unit of Analysis	Racial or Income Disparities/ Number of Studies	Racial Disparities/ Number of Tests	Income Disparities/ Number of Tests	Race More Important than Income/ Number of Tests
Nation (random samples)	13/13/ (100%)	9/9 (100%)	4/5 (80%)	1/1 (100%)
States	1/1 (100%)	1/1 (100%)		
State (random samples)	3/3 (100%)	3/3 (100%)	0/1 (0%)	1/1 (100%)
Urban areas	4/4 (100%)	4/4 (100%)	3/3 (100%)	2/2 (100%)
Counties	8/8 (100%)	10/12 (83%)	6/13 (46%)	6/6 (100%)
City (random samples)	5/5 (100%)	4/4 (100%)	1/1 (100%)	
Census places	1/1 (100%)	1/1 (100%)	0/1 (0%)	1/1 (100%)
Towns	4/4 (100%)	7/10 (70%)	6/9 (67%)	2/5 (40%)
5-digit ZIP code areas	7/8 (88%)	6/8 (75%)	6/7 (86%)	5/5 (100%)
Census tracts	18/18 (100%)	13/15 (87%)	16/17 (94%)	4/9 (44%)
TOTAL	63/64 (98%)	59/68 (87%)	43/58 (74%)	22/30 (73%)

Reprinted by permission of the Corporate Consumer Council of the National Wildlife Federation from Goldman 1993.
* Some of the studies and tests involved more than one environmental concern, so the numbers in each column do not necessarily add up to totals in the bottom row. Blanks indicate tests were not performed.

the auspices of the largest waste management firm in the world, WMX Technologies, found no difference in the environmental hazards to which various income and racial groups are exposed. Goldman points out, however, that "WMX may have a corporate interest in refuting the findings of disproportionate impacts from commercial waste management because of the negative publicity that such findings have generated" (Goldman 1993). Indeed, Goldman's own reexamination of the WMX study finds that its final report did not accurately reflect its own data. "WMX's own numbers," he continues, "reveal that more than a half-million nonwhites live in ZIP code areas with WMX waste facilities, comprising a 35 percent greater share of their total population than the average for the country as a whole."

The other major conclusion to be drawn from studies of environmental inequities is that race is a better predictor of inequity than is income. This point has been an issue of dispute from the earliest days of the environmental justice movement. Are poor white people just as likely to be exposed to environmental insults as are poor nonwhites? Do polluters take advantage of communities because they are low-income communities or because they consist primarily of minorities?

Some early studies did not examine this question. But among those that did, the answer was clear: In three-quarters (22 of 30) of the studies in which this question was tested, race rather than income was the more important factor. As Table 3 shows, this conclusion is born out at almost every geographical level, from state and national to local.

A number of researchers who have explored the correlation of both race and income with environmental inequities have come to the same conclusion as has Goldman. In a review of 16 studies by Paul Mohai and Bunyan Bryant, for example, the authors conclude that "[the review] provide[s] clean and unequivocal evidence that income and racial biases in the distribution of environmental hazards exist. Our findings also appear to support the claims of those who have argued that race is more importantly related to the distribution of these hazards than income" (Mohai and Bryant 1992).

Differences of Opinion

In spite of all that has been said above, there remains a doubt in some people's minds as to whether environmental inequities actually exist. The question has been raised as to whether the evidence summarized above may reflect research methodologies rather than real differences in exposure to pollutants among various economic classes and races.

One study exploring this issue was funded by Chemical Waste Management, one of the world's largest waste management companies. The study was led by Douglas L. Anderson of the University of Massachusetts at Amherst. Anderson's research team took the approach of focusing on the smallest population units for which reliable national data are available. These units are the census tracts used by the U.S. Bureau of the Census in its decennial census.

When the Anderson team compared the demographic characteristics of census tracts containing toxic waste treatment, storage, and disposal facilities (TSDFs) with similar tracts lacking such facilities, they found essentially no differences in the racial status of residents of the two tracts. Their summary of this research included three major conclusions:

> First, the appearance of equity in the location of TSDFs depends heavily on how areas of potential impact or interest are defined. Second, using census tract areas, TSDFs are no more likely to be located in tracts with higher percentages of African Americans and Hispanics than in other tracts. Third, the most significant and consistent effect on TSDF location of those we considered is that TSDFs are located in areas with larger proportions of workers employed in industrial activities, a finding that is consistent with a plausibly rational motivation to locate near other industrial facilities or markets (Anderson et al. 1994).

Some of the strongest doubts about the existence of environmental inequities have come from industries that have themselves been accused of producing such effects. The case of WMX Technologies, cited above, is an example. Another example is that of the textile industry that for some years denied any connection between cotton dust, a common environmental problem in the industry, and the disease known as "brown lung" or byssinosis.

Government officials have also been known to adopt an industry view of environmental inequities, or at least to find reasons to excuse such patterns. In response to the report by the Louisiana Advisory Committee (LAC) to the U.S. Commission on Civil Rights, Gary Johnson, of the Louisiana Department of Environmental Quality, pointed out the dangers of presenting too dismal a view of the state's environmental problems.

> I feel that if the data is published as presented [in the LAC report] this would place a negative connotation on the state of Louisiana regarding our on-going efforts toward toxics reduction in [the] Baton Rouge-New Orleans corridor . . . I want to clearly point out that how regulatory agencies present data in the future will clearly impact the cooperation we receive in return from industry and industry association . . . Anyone working in the environmental field today should be extremely cautious in publishing information regarding "environmental equity" and specific geographical regions, specifically the Lower Mississippi Corridor (as reported in Bullard 1994a).

The Origins of Environmental Inequities

At the conclusion of their review of 16 major studies on the relationship between race and income and the distribution of environmental hazards, Paul Mohai and Bunyan Bryant, faculty members at the University of Michigan School of Natural Resources and organizers of the Michigan Conference on Environmental Justice, raise an interesting point. "Ultimately, knowing which [race or income] is more important may be less relevant, however, than understanding the conditions associated with race and class that appear to consistently, if not inevitably, lead to inequitable exposure to environmental hazards and in understanding how these conditions can be addressed and how inequities in the distribution of environmental quality can be remedied" (Mohai and Bryant 1992).

Precisely how environmental inequities develop is a complex question, one that has not yet been answered to everyone's complete satisfaction. Yet, some forces contributing to this state of affairs are reasonably clear.

Progress and Environmental Degradation

In the first place, the issue of environmental degradation comes about because societies strive for a better way of life—nicer, more comfortable homes, better clothes, more effective medicines, and more efficient means of transportation, for example. The largest single difference between the "have" and "have-not" countries of the world reflects the abilities of some nations to provide their citizens with better material goods and a larger supply of energy—a higher standard of living—than can other nations.

But the systems of production, distribution, and consumption of material goods and energies always result in the release of waste products into the environment. Polluted air and water, hazardous waste dumps, and dangerous occupations are all examples of the price that a society pays for improving its standard of living. And the more the standard of living is improved, the greater the amount of waste released to the environment.

It might be that in a perfect world, the benefits and burdens of a high standard of living might be distributed equally throughout a society. Everyone would have equal access to new products and would have to live with increased levels of environmental degradation. But no perfect society exists. Instead, in every form of government devised—socialism, communism, fascism, or democracy—there are "haves" and "have-nots." Some people receive a larger share of the benefits, and others receive a larger share of the burdens. One of the goals of almost any social movement is to obtain a more fair distribution of these benefits and burdens.

How is the distribution of benefits and burdens determined in the real world? One factor is the way in which various groups within the society are viewed. In the United States (as in many other nations), white people tend to have the advantage over people of color, men over women, the physically well over the physically handicapped, heterosexuals over homosexuals, and so on. In these nations then, white males tend to receive the highest proportion of material goods available in the society. And the burdens that develop as a result of an improved lifestyle for the more privileged are likely to fall on poor people of color. The environmental justice movement has fought to make that fact abundantly and unquestionably clear.

NIMBY and PIBBY

Beyond political and social philosophy, however, there are some concrete reasons that the burdens of environmental degradation have fallen disproportionately on communities of color and on people of low economic status. In the first place, decisions as to where polluting industries, hazardous waste dumps, radioactive storage sites, and other environmentally undesirable sites are to be located are commonly made by governmental bodies, such as city councils, planning departments, or zoning committees. It is not uncommon for such bodies to consist of a disproportionate number of the local "power structure," which usually means white males. The tendency may be in many cases, then, for environmentally undesirable sites to be located in "their" neighborhoods rather than "our" neighborhoods.

This tendency is reinforced by the reluctance of ordinary citizens to have such sites located in their neighborhoods. One of the most familiar phrases in land use issues in the last few decades has become "not in my backyard," or NIMBY. The phrase reflects the fact that most people recognize that undesirable factories, prisons, halfway houses, hazardous waste sites, and other "locally unwanted land uses" (LULUs) are now a reality in our society. Given our way of life, they have to exist. But few people want them near their homes. They would prefer to have them constructed "somewhere else." In many cases, this has led to a new attitude, the "put it in the blacks' backyard," or PIBBY syndrome.

Concerns about Environmental Issues among People of Color

Consonant with the NIMBY principle has been the belief, spoken or unspoken, that African Americans and other minorities are not concerned about environmental issues. They have too many other concerns—jobs, housing, and health problems, for example—to care about the surroundings in which they live, goes this argument. This attitude is clearly racist and offensive, suggesting that African Americans and other minorities somehow lack the same appreciation as whites for the health and beauty of their environment. The attitude has now been shown also to be factually incorrect.

In his study of environmental problems in five largely African American communities, Robert Bullard found that residents were very much interested in and concerned about such issues. Overall, more than half of the respondents in Bullard's survey took part in some form of environmental activism, such as writing a letter or telephoning an official about an issue, signing or circulating a petition, attending or organizing a meeting in one's home, marching in a demonstration, attending a public meeting, or helping to raise funds (Bullard 1994a; also see Buttel 1987; Lowe and Pinhey 1982; and Taylor 1989).

An associated finding of the Bullard study was that environmental activism among African Americans tended to take place not within traditional environmental groups, but through other organizations, such as churches, social clubs, labor unions, civil rights groups, or parental groups. On average, 16.3 percent of Bullard's subjects belonged to an environmental group of some kind compared to memberships of 76.5 percent in churches, 27.9 percent in community involvement groups, and 23.1 percent in parent groups (Bullard 1994a).

Access to Tools of Protest

The most important single reason that LULUs end up in neighborhoods with a majority of people of color or of people in lower economic strata may be the difference in access that various groups of people in the United States have to the political process. One might argue that a community of color has the same opportunity to protest the siting of a hazardous waste site or a polluting industry as does any other community. It can take its case before a zoning board or bring suit in a court of law to prevent the siting of an undesirable facility, for example.

The reality is, however, that people of color and poor people probably lack the experience, the training, or the financial resources to engage in the same kinds of battles that affluent white communities routinely use to keep their neighborhoods free of environmentally offensive sites. Mohai and Bryant have summarized the disadvantages under which communities of color operate when confronted with environmental hazards.

> These communities tend to be where residents are unaware of the policy decisions affecting them and are unorganized and lack resources for taking political action; such resources include time, money, contacts, knowledge of the political system, and others. Minority communities are at a disadvantage not only in terms of availability of resources but also because of underrepresentation on governing bodies when location decisions are made. Underrepresentation translates into limited access to policy makers and lack of advocates for minority interests (Mohai and Bryant 1992; citations omitted).

Job Blackmail

The siting of LULUs in communities of color or low-income communities often presents a difficult dilemma. The construction of a factory or a hazardous waste site may mean new jobs and an expanded tax base for such communities, an important economic incentive when unemployment may be high and municipal services minimal. When Chemical Waste Management, Inc. (CWM), purchased an existing hazardous waste landfill in Sumter County, Alabama, in 1978, for example, it worked hard to convince local residents of the landfill's economic benefit to the community. A chart it made public, "How Much Will We Get from Chem Waste?" showed that 21 different governmental and nongovernmental bodies, from the Sumter County General Fund to the Sumter County Board of Education; Sumter County Water Authority; Livingstone University General Fund; towns of Gainesville, Geiger, and Emelle; cities of Livingstone and York; and the Sumter County Historical and Preservation Society, the Sumter County Library System, and the Sumter County Fine Arts Council all receiving tax benefits from the landfill. Overall, CWM was to become the largest single taxpayer in the county, producing about half of all public funds collected (Bailey and Faupel 1992).

Poor communities find it difficult to reject an economically productive facility such as this one, even if it might present serious environmental and health hazards. Individual workers face a similar dilemma. It seems unlikely that anyone would choose to work in a hazardous waste landfill or a factory in which noxious fumes are constantly released, but a dangerous job may be better than no job at all.

This form of "job blackmail" is hardly new to minority workers. Throughout history, African Americans, Asian Americans, Hispanic Americans, or other minority groups have been given jobs that whites would not want or take. Migrant workers probably do not choose to work 12-hour days in hot, dry fields where they are exposed to dangerous pesticides, for example, because they enjoy the outdoor life. They take such jobs because they may be the only ones open to them.

In many cases, the assignment of minorities to dangerous and difficult jobs has been justified by pseudo-scientific theories about differences in physical or biological traits among various ethnic groups. For many years, for example, the iron and steel industry assigned African American men to work at its coke ovens because of the supposed ability of African Americans to withstand heat better than whites. Similarly, workers with dark skin were once more frequently assigned to work with caustic chemicals in some industries because job-induced skin irritations would supposedly be less noticeable and less objectionable than they would be with lighter skinned workers (Davis 1977).

Job blackmail and economic blackmail present, therefore, a damned-if-you-do and damned-if-you-don't choice for both individuals and communities. As Robert Bullard has observed,

> Industries such as paper mills, waste disposal and treatment facilities, heavy metals operations, and chemical plants, searching for operating space, found minority communities to be a logical choice for their expansion. These communities and their leaders were seen as having a Third World view of development. That is, "any development is better than no development at all." The sight and smell of paper mills, waste treatment and disposal facilities, incinerators, chemical plants, and other industrial operations were promoted as trade-offs for having jobs near "poverty pockets" (Bullard 1992).

The Effects of Environmental Regulations

The picture painted thus far of the environmental inequities faced by people of color and poor people seems bleak. Yet, the United States has developed a much more enlightened view of the environment in the past three decades. People in general are probably much better informed and more concerned about the environment, and federal, state, and local governments have become much more aggressive about passing legislation and enforcing regulations that will protect the environment. Do these changes suggest that progress is being made in dealing with the disproportionate exposure faced by people of color and poor people in the United States?

It would be difficult to argue that no overall improvement has been made as a result of such changes. Somewhat remarkably, however, it is also clear that many new environmental regulations have actually increased the environmental hazards to which minorities are exposed. Again, the benefits of environmental regulations for some groups of people have resulted in an increased burden for other groups of people.

In the first place, it is clear that new programs to protect the environment cost money. Those costs are borne by all taxpayers in the society, poor as well as rich. They are borne by all taxpayers whether or not they benefit directly from the environmental program in question. Programs to protect national wildlife areas or national seashores are paid for by African Americans, Hispanic Americans, and Native Americas, as well as by whites. Every time an environmental regulation raises the cost of an automobile because of new pollution control devices, raises the cost of housing because of new land use regulations, or raises the costs of food products because of new pesticide-use rules, people belonging to lower income groups pay a disproportionate percentage of that cost.

More to the point, the economic costs of new environmental regulations are more likely to have a more significant effect on low-income people than on high-income people. If the costs of environmental regulations average out to $500 per person nationwide, those costs will be more detrimental for someone making $10,000 per year than for someone making $100,000 per year.

One attempt to measure the economic impact of the air pollution policy in the United States is that of Leonard P. Gianessi, Henry M. Peskin, and Edward Wolff. This study was used to determine the costs and benefits that could be anticipated from the full implementation of the federal Clean Air Act. The researchers found that the effects of the Clean Air Act tend to be regressive; that is, that people of low income tend to pay more proportionally than do those of high income. For those with an income in the range between $3,000 and $8,000 per year, the net costs of the Clean Air Act were about 0.90 to 1.00 percent of their income. In contrast, those with incomes of more than $8,000 uniformly experienced a net cost of about 0.50 percent of their income (Gianessi, Peskin, and Wolff 1979; also see Johnson 1980).

The costs of environmental regulation can also be measured in terms of exposure to regulated pollutants. In a master's thesis titled *The Distribution of Outdoor Air Pollution by Income and Race: 1970–1986*, Michel Gelobter found that all races and all income groups in selected urban areas were exposed to lower levels of ambient air pollutants at the end of that period than they were at the beginning. However, Gelobter found differences in exposure among both income and racial groups. For example, in 1970, 1975, 1980, and 1984, the average white resident of the urban areas studied was exposed to 100, 75, 75, and 62 micrograms of total suspended particulates per cubic meter of air, respectively, while the comparable figures for nonwhites during those four years were 122, 85, 82, and 70 micrograms of total suspended particulates per cubic meter of air, respectively. In a more sophisticated analysis of the "relative benefit from air quality improvements," including a variety of measures, Gelobter found that whites and nonwhites rated almost exactly the same in 1970. Over the next 15 years, however, the difference in "relative benefits" diverged until whites were approximately 5 percent "better off" on this measure (Gelobter 1992).

A number of studies have produced another interesting finding: The U.S. Environmental Protection Agency (EPA), the U.S. government agency primarily responsible for the prevention and clean-up of polluted sites, has itself been guilty of discriminatory policies and practices in dealing with communities of color and low economic status. In a study conducted for Clean Sites, Inc., in 1990, for example, Kate Probst found that the EPA was less likely to include rural poor communities on its list of potential Superfund sites than other communities (Probst 1990).

One of the most telling studies on this issue was that conducted by Lavelle and Coyle in 1992. These researchers found, first of all, that the EPA was likely to make minority communities wait even longer than low-income communities to receive Superfund listing. The pattern was for low-income areas to wait 11 percent longer than high-income areas, and for minority areas to wait 20 percent longer than white areas. Lavelle and Coyle also found that once a minority community did receive a Superfund listing, the agency was more likely to institute a program of waste containment than to arrange for toxicity reduction or removal, as was the case with white communities. Finally, the fines assessed by EPA averaged only 20 percent for instances of pollution in minority communities compared to those assessed in white communities (Lavelle and Coyle 1993).

Critics have argued that an indication of the EPA's attitudes toward racial issues can be found in the agency's hiring policies. That is, one way in which the agency could demonstrate a greater sensitivity to issues of environmental inequities, they suggest, would be to make greater efforts to include members of minorities on its staff. Yet, according to an agency report published in 1992, less than 10 percent (33 out of 412) of new management positions filled in the previous year were given to members of a minority. In the following year (1992), 42 of 354 management hires went to minorities. In comparison, the number of white females hired during these two years were 142 (34 percent of new hires) and 126 (36 percent) in 1991 and 1992, respectively (U.S. Environmental Protection Agency 1992).

In a critique of the nation's environmental protection policies, Robert Bullard has written that these policies have

> (1) institutionalized unequal enforcement; (2) traded human health for profit; (3) placed the burden of proof on the "victims" and not on the polluting industry; (4) legitimated human exposure to harmful chemicals, pesticides, and hazardous substances; (5) promoted risky technologies, such as incinerators; (6) exploited the vulnerability of economically and politically disenfranchised communities; (7) subsidized ecological destruction; (8) created an industry around risk assessment; (9) delayed cleanup actions; and (10) failed to develop pollution prevention as the overarching and dominant strategy (Bullard 1994a).

Environmental Inequities: By Chance or By Choice?

Even if everything that has been said above were entirely true, one critical question remains, the question of intent. Do environmental inequities exist because corporations, governmental bodies, and/or individuals plan for them, or do they come about simply as the result of business and/or technical decisions that are based on other factors? This question has been at the core of environmental justice issues because from the earliest days of the movement, the assumption has been that environmental inequities are a manifestation of racism and, perhaps to a lesser degree, class issues. Recall that Benjamin Chavis defined environmental racism as "the deliberate targeting of people of color communities for toxic waste facilities and the official sanctioning of a life threatening presence of poisons and pollutants in people of color communities."

Some people, however, while acknowledging the existence of environmental inequities, argue that racism is a minor or nonexistent factor. Such inequities come about not because business, industry, and government intend to expose poor and minority communities to environmental insults, these people claim. Instead, these problems arise as a by-product—and an unfortunate by-product that needs to be addressed—of business decisions that have no racial or class component.

This position was expressed in a letter to Chavis from J. Winston Porter, then Assistant Administrator for Solid and Hazardous Wastes for the EPA. "There's no sociology to it [siting decisions]," Porter wrote. "It's strictly technical" (as cited in Lee 1992).

This issue is, of course, a crucial one within the environmental justice issue. On the one hand, it may be that corporations and governments routinely make decisions about the siting of LULUs based strictly on economic factors, such as labor costs, availability of land, and access to transportation. Racial, ethnic, or other social factors might have little or no impact on such decisions. In such a case, minority communities could expect to work with corporations and governments to find ways of siting such facilities with less disproportionate impact on themselves. The affected communities and the agencies responsible for environmental inequities could work together with trust and confidence in each other.

On the other hand, it might be possible that factors other than economic matters enter into the siting of LULUs. One could imagine that those responsible for making such decisions hold African Americans, Hispanic Americans, Asian Americans, Native Americans, and poor people in less regard than they do middle- and upper-class white citizens. In such a case, solving the problem of environmental inequities would pose a different dilemma. Minority communities would be faced with the task of changing attitudes among those in corporations or government whose respect they currently do not have. Or, they might have to use a more confrontational approach to problem-solving (such as court cases) to deal with environmental inequities. As a matter of fact, it is the latter position that most of those active in the environmental justice movement appear to hold and to have held since the earliest days of the movement.

Still, many observers continue to deny the presence of social attitudes, such as racism, in the origin of environmental inequities. An example is the comment made by Kent Jeffreys, Director of Environmental Studies for the Competitive Enterprise Institute, in testimony before the Subcommittee on Civil and Constitutional Rights of the U.S. House of Representatives on 3 March 1993. Jeffreys said that

> Racism exists. Environmental problems exist. These facts, however, do not reveal whether or not *environmental racism* is occurring. Regardless of whether any particular case fits the definition of environmental

racism, the fact remains that environmental problems—from a minority perspective—are rather trivial
in comparison to the larger economic and civil liberty
issues: solve these and you have solved most, if not all,
of the environmental inequities (U.S. Congress 1993).

This viewpoint was reflected during hearings before the
Louisiana Advisory Committee to the U.S. Commission on Civil
Rights in February of 1992. The hearings were held to determine
"the status of environmental problems in selected areas of the
State." At the conclusion of its fact-finding sessions, the committee concluded that "many black communities located along the
industrial corridor between Baton Rouge and New Orleans are
disproportionately impacted by the present State and local government system for permitting and expansion of hazardous
waste and chemical facilities." The committee did not specifically
attribute the environmental inequities it found to racism, but its
final report contains a number of allusions that suggest that the
majority of committee members may have held this viewpoint.
At one point the report says, for example, that

> . . . some black citizens and organizations view with
> skepticism and distrust some agencies in State gov-
> ernment, including the Department of Environmental
> Quality. These views are held because of poor access
> to government and an ongoing perception that State
> government discriminated against them to promote
> and sustain the interests of industry and business
> (Louisiana Advisory Committee 1993).

Perhaps the best view of the committee's mindset on this
issue comes from one member who wrote a dissenting statement
to the report. John S. Baker, Jr., Professor of Law at the Louisiana
State University Law Center, concluded for himself that "Absent
from the report is the one finding most clearly supported by the
evidence: *Environmental Racism has not been shown to exist in
Louisiana* . . . None of the extensive findings contains anything
about, nor could they support a finding of, 'deliberate targeting'
or 'official sanctioning'" [emphasis in original]. Baker argued
that members of the committee knew in advance of the hearings
that discriminatory motive would not be found, and that they
were conducted with other purposes in mind than the search for
such a motive. Baker continued in his dissent:

The overall handling of this project reminds me of the situation that occasionally occurs during a criminal case. Sometimes during voir dire when a potential juror is asked whether he or she can presume the innocence of the defendant and give him a fair trial, the juror will answer quite innocently in terms that effectively say: "Of course, I'll give him a fair trial before convicting him." Like the prospective juror, the staff and committee members, I believe, are acting with the best of intentions. Nonetheless, the report, and the process which produced it, have blindly stepped over and around the evidence which so clearly establishes no racially motivated discrimination on environmental decisions in Louisiana (Louisiana Advisory Committee 1993).

The issue being raised in this debate might be characterized as a chicken-and-egg debate: which came first, the hazardous waste facility or low-income and minority communities? That is, do industries tend to choose an existing minority and/or low-income community in which to locate because it has less regard for the health and welfare of people living there, or does an industry build a landfill (or other LULU) and low-income and minority people then choose to move into that area?

Some research has been done on this question, but the findings of that research are not yet conclusive. For example, Vicki Been, Associate Professor of Law at New York University School of Law, has reanalyzed the demographics of waste disposal sites previously studied by the GAO and by Robert Bullard in Houston. Her thesis was that greater attention should be paid to the economic and racial status of an area before a toxic waste landfill was constructed than at the current time.

Using this criterion, Been obtained somewhat conflicting results from the GAO and Bullard studies. In the former case, she found that the four communities where waste disposal sites were sited were originally both predominantly low-income and minority communities. She also found that neither the proportion of low-income nor the proportion of minority individuals increased after the siting. In the Houston case, Been found that waste disposal sites were occupied predominantly by minority populations, but not by low-income populations, before facilities were installed. After the facilities were opened, however, the percentage of African Americans and of low-income people increased.

Been concludes from her research that

> . . . research examining the socioeconomic characteristics of host neighborhoods at the time they were selected, then tracing changes in those characteristics following the siting, would go a long way toward answering the question of which came first—the LULU or its minority or poor neighbors. Until that research is complete, proposed "solutions" to the problem of disproportionate siting run a substantial risk of missing the mark (Been 1994).

One of the difficulties in conducting studies of this type is the difference between older hazardous waste sites, many now closed, and new or proposed sites. At one time, it was relatively easy to build a hazardous waste site, and such older sites can be found throughout the United States. Now, obtaining permission to construct a hazardous waste site is much more difficult. It follows that the decision as to where to build a newer site might, therefore, be somewhat different than it was in the past.

One study bearing on this issue is the research of James T. Hamilton. Hamilton found that counties planning to expand waste disposal sites apparently made no distinction between areas on the basis of racial or economic conditions. He also found, however, that sites scheduled for closure were more likely to be located in white areas than in minority communities (Hamilton 1993).

In his summary of new research needed in the field of environmental justice, Benjamin Goldman returns to the question of causality in the siting of LULUs. "Since the majority of the studies reviewed here use correlational analysis, and focus on policy impacts rather than procedures," he writes, "they do not prove definitively that the observed disparities are the result of any consistent causal mechanism" (Goldman 1993).

Yet, as Goldman also points out, studies like those of Been and Hamilton deal with the very core of the environmental justice movement. He goes on to say that

> So the question raised by the environmental justice movement is how do the decisions of legal, cultural, economic, and other social institutions yield such disproportionate patterns of environmental impact? Further, what legal models are available to achieve

redress for harm without proving intent? Environmental justice activists suggest that environmental racism must be examined within the broader context of racism in society; and that further research and action focusing on these complex issues are as legitimate as those involving the many other forms of discrimination found in housing, employment, education, etc. (Goldman 1993).

The Legal Question of "Intent" in Environmental Inequities

Goldman's summary refers to another key issue in the dispute over environmental inequities: the question of intent. Many involved in the environmental justice movement would like to demonstrate that environmental harm has come as a result of decisions and actions taken by business, industry, or governmental bodies and then to obtain some relief from that harm from the appropriate body. That is, an environmental justice organization might like to prove that Company X placed its hazardous waste landfill adjacent to a predominantly African American community because that was the easiest decision for the company to make. The organization might then argue that health effects to the community resulting from the company's decision were the company's responsibility, and that the company should be forced to pay a financial penalty for its decision.

And in some cases, this approach has worked, and companies have paid large fines or made large financial settlements to members of communities harmed by their activities. But such instances are relatively rare. One reason is that a stringent standard of proof is often required by courts in cases of environmental harm. This standard was established in the landmark decision of *Washington, Mayor of Washington, D.C., et al. v. Davis et al.* In that case, the court ruled that a plaintiff must be able to prove that harmful actions taken by an individual, a group, or a corporation were intended to cause harm to the plaintiff and not that the harm occurred as an unexpected by-product of the action. This legal standard is a difficult one to employ since it requires that a plaintiff somehow find out what was in the minds of a businessman or government official and then demonstrate "intent" to the court. As one legal observer has written,

. . . the practical effect of the required "discriminatory intent" element is devastating to most civil rights claims because of the inordinate difficulty of proving the subjective, motivating intent of a decisionmaker (Lazarus 1993; also see Lavelle and Coyle 1993).

The hurdles created by the "intent" standard have meant that many environmental justice activists have used means other than legal challenges to deal with environmental issues. As one example, the Lumbee Indians of Robeson County, North Carolina, were involved in an extended dispute with the GSX Corporation, which planned to build a hazardous waste facility on lands sacred to the Lumbee. The tactics used by the Lumbee were not legal challenges, but a more personal campaign featuring traditional Native American dance and music at public hearings, leafleting at churches, and participation in all planning sessions held on the GSX request (Austin and Schill 1994). Grassroots activities such as these have been common and successful in situations when long, expensive, and complex legal challenges would have held much less promise.

Responding to Environmental Inequities

Some obvious methods for dealing with environmental inequities have been suggested thus far. Some communities have found that direct political action—protests and demonstrations, for example—have accomplished their goals. Others have sought relief in the courts. In general, a debate remains as to the best mechanism by which a community dealing with a LULU can protect itself against the dangers of such a facility.

On the one hand, some people believe that the question of environmental justice is, to a large extent, a problem with which governments have to deal. New and better laws and stronger enforcement of existing laws can prevent future environmental inequities and can help deal with such instances now existing. Robert Bullard, for example, has called for government to take five steps to ensure environmental justice: national legislation modeled on civil rights acts; the elimination of existing environmental hazards; shifting the burden of proof for guilt or innocence in environmental inequity cases from affected communities to polluting industries; modifying the current legal standard of "intent" in dealing with cases of environmental inequities; and providing compensation and assistance to communities most seriously affected by environmental inequities (Bullard 1994c).

An opposing view to the solution of environmental inequities calls for greater dependence on market forces in the private sector, and less on government involvement. Companies that benefit from the use of land for the construction of LULUs should be required to pay a fee for this privilege. That fee should then be transferred to minority and poor communities on whom the burdens of the LULU siting fall.

An example of this approach is legislation proposed in the State of Wisconsin in 1981. This legislation required negotiations between a company wishing to build a hazardous waste facility and the municipality in which the facility would be located. Three advantages of the Wisconsin plan, as noted by Christopher Boerner and Thomas Lambert of the Center for the Study of American Business at Washington University in St. Louis, are that the statute

> . . . clearly specifies "the players of the game"—that is, who negotiates with whom. Both developers and potential host communities are required to establish negotiating committees, with the rules regarding these representatives explicitly set forth in the statute. Secondly, the legislation assures that these "players" will not only negotiate, but also that the results of their negotiations will be legally binding. . . . Finally, the Wisconsin statute provides a "back-up plan"—a way to arbitrate siting decisions should negotiations fail or should one party refuse to cooperate (Boerner and Lambert 1995).

The question of how best to deal with environmental inequities requires some special considerations in issues involving Native Americans. The placement of environmentally harmful facilities on Native American lands has been especially common in the last few decades. One reason for this pattern is that many of the lands encompassed by Indian reservations or otherwise owned by tribes are especially rich in minerals, such as coal, oil, and uranium. In order to extract those minerals, companies have to deal with tribal governments. In the second place, many Native Americans have relatively modest incomes, and the opportunity to improve their economic status by leasing lands to mining companies is tempting.

The controversy that has developed is between those who feel that the siting of mines and processing facilities is just one more form of environmental inequity (and, probably, of environmental racism) that must be protested and prevented and those who feel that some level of environmental degradation is acceptable provided that adequate financial compensation is provided (Small 1994; Skull Valley Goshute Tribe Executive Office 1993).

Environmental Justice as an International Issue

For many of those on the front lines of the environmental justice movement in its early stages, international issues were probably of little or no consequence. By its very nature, the movement was usually concerned with environmental issues at the neighborhood, town, or county level. And yet, trade policies and practices of the United States government and of many multinational corporations had reflected a pattern of environmental inequity long before that term had ever been coined.

One of the important but unexpected consequences of the American environmental movement of the 1960s and 1970s was that the United States began to adopt laws and regulations far more severe than those of most other nations of the world, certainly more restrictive than those of less developed nations. Pesticides that could no longer be used in the United States because of their toxicity were, in many cases, still legal in other nations. Companies who could no longer sell their products to American farmers because of their toxicity might well be able to sell them to farmers in Indonesia, Ghana, Somalia, and other struggling nations (Norris 1982).

Such practices could be thought of as just another form of environmental inequity, and perhaps environmental racism, but on a much grander scale. During the late 1980s and early 1990s, a number of waste disposal problems made this analogy even more obvious. In 1988, for example, the city of Philadelphia hired a Norwegian company, Bulkhandling, Inc., to transport 15,000 tons of toxic incinerator ash to Kassa Island in the African nation of Guinea. Only when plant and animal life began to die off on the island did the government discover what had happened and ordered Bulkhandling to remove and return the ash to Philadelphia (Mpanya 1992).

As permits and land for hazardous waste landfills on the continental United States became more and more difficult to obtain, dumping in African nations became more and more attractive. The number of such cases increased by a factor of 10 in the three-year period between 1987 and 1989, rising from 3 cases in 3 countries in the first of these years to 40 cases in 25 nations in 1989 (Mpanya 1992).

As with possible cases of environmental racism in the United States, such practices often make good economic sense: they provide an answer to a difficult corporate problem (waste disposal) while bringing significant economic benefit to people who need it (poor African nations). In fact, this argument in a slightly different context was made by Lawrence Summers, Chief Economist of the World Bank, in a memo of 12 December 1991. The memo was obtained by and then reprinted in the British journal, *The Economist*, in its 8 February 1992 issue. "Just between you and me," Summers wrote, "shouldn't the World Bank be encouraging *more* migration of the dirty industries to the LDCs [less developed countries]?"

Summers presented three reasons for his recommendation. For example, he said, people who live in a country where the life expectancy is fairly short are probably not going to worry very much about getting prostate cancer late in life. They are more likely to want to improve the quality of life for the years that they do have. Considering arguments such as this one, Summers came to the conclusion that "the economic logic behind dumping a load of toxic waste in the lowest-wage country is impeccable and we should face up to that" (Foster 1993).

A similar view was expressed by Harvey Alter, Manager of the Resources Policy Department of the United States Chamber of Commerce. Humanitarian instincts prompt us to want to cut back on our trade in hazardous wastes, Alter agrees, but the long-term effect will be disastrous for underdeveloped countries. The net effect of such efforts, he says, "will be the deepening poverty of the developing nations, a poverty that will exacerbate the environmental problems these actions were trying to alleviate" (Alter 1995).

Still, many of the objections raised by the environmental justice movement to environmental inequities at the local level have relevance on an international scale. The notion that the benefits resulting from a higher standard of living should accrue to some groups of people and the burdens should be relegated to another group is as valid whether the groups are racial or economic classes within the United States or nations at various stages of development around the world.

One recognition of this fact is reflected in the adoption in 1989 of the Basel Convention on the Control of Transboundary Movements of Hazardous Wastes and Their Disposal. More than 100 nations, including the United States, have now ratified or indicated their intention to ratify this treaty. The treaty requires that a company wishing to ship hazardous wastes from one nation to another must first notify the government of the receiving country and obtain written permission to do so. Any country not wishing to accept a shipment of hazardous waste may decline to give its permission (Basel Convention on the Export of Waste 1991).

The issue of environmental inequities on an international level has introduced another factor into the environmental justice movement in the United States. An African American, Hispanic American, Native American, or Asia American community may very well—and rightfully so—fight to prevent the siting of a hazardous waste dump in its own community. Its objections to being exposed to the health hazards of such a facility are perfectly valid.

But in conducting this kind of battle, communities of color must also consider the question of where such sites are to be located. If Corporation XYZ cannot build a hazardous waste dump in Alabama, will it try, instead, to ship those wastes to some African nation? After all, as long as we maintain our current standard of living, hazardous wastes will be produced and a way of disposing of them will have to be found.

The Future of Environmental Justice

The fact that such dilemmas now arise is one of the exciting and, perhaps, unexpected consequences of the environmental justice movement. The question that may have to be asked in the future is not simply where to put the hazardous wastes and dangerous industries that our booming economy has spawned. Instead, it may be whether we really want and need to have the social, political, and economic system in which such issues have to arise. One of the questions raised by the environmental justice movement on its most fundamental level, then, is not just the distribution of environmental inequities, but the necessity for such inequities at their current level in the first place.

Indeed, it is somewhat ironic that a movement that began as and to a large extent continues to be a decentralized, local movement has raised some of the most profound questions about the structure and dynamics of American society. Those involved in the environmental justice movement have pointed out how much their experiences have raised questions about class structure, gender roles, economic opportunity, political power, and other fundamental issues in American society. A writer in *Everyone's Backyard*, the newsletter of the Citizens Clearinghouse for Hazardous Wastes, has written that

> Environmental justice is a people-oriented way of addressing "environmentalism" that adds a vital social, economic and political element . . . the new Grassroots Environmental Justice Movement seeks common ground with low-income and minority communities, with organized workers, with churches and with all others who stand for freedom and equality. . . . Environmental justice is broader than just preserving the environment. When we fight for environmental justice, we fight for our homes and families and struggle to end economic, social and political domination by the strong and greedy (as quoted in Szasz 1994).

Attitudes such as these suggest that the environmental justice movement may be the core around which an even broader civil rights movement may form. Gary Cohen of the National Toxics Campaign has predicted that

> We see this as . . . the civil rights movement of the nineties. This is the antiwar movement of the nineties, and this is an area of tremendous potential social change. . . . The toxics issue is a very powerful, powerful issue to organize people for social change. We feel that it is time to really start being more self-conscious about that (as quoted in Szasz 1994).

The protestors at Warren County, North Carolina, in 1983 probably never had a vision such as this one when they fought against a PCB landfill in their backyard. But they might well have been very proud of the ultimate direction of the movement for which they were so largely responsible.

References

Alter, Harvey. 1995. "Halting the Trade in Recyclable Wastes Will Hurt Developing Countries." In *Environmental Justice*, edited by J.S. Petrikin. San Diego: Greenhaven Press (Chapter 9).

Anderson, Douglas L., et al. 1994. "Hazardous Waste Facilities: 'Environmental Equity' Issues in Metropolitan Areas." *Evaluation Review* 18 (2): 123–140.

Austin, Regina, and Michael Schill. 1994. "Black, Brown, Red, and Poisoned." *The Humanist*, (July/August): 9–16.

Bailey, Conner, and Charles E. Faupel. 1992. "Environmentalism and Civil Rights in Sumter County, Alabama." In *Race and the Incidence of Environmental Hazards: A Time for Discourse*, edited by Bunyan Bryant and Paul Mohai. Boulder, CO: Westview Press (Chapter 11).

Basal Convention on the Export of Waste. 1991. Hearing before the subcommittee on Transportation and Hazardous Materials of the House Committee on Energy and Commerce, 102nd Congress, 1st Session, 10 October 1991.

Been, Vicki. 1994. "Locally Undesirable Land Uses in Minority Neighborhoods: Disproportionate Siting of Market Dynamics?" *The Yale Law Journal* 103: 1383–1422.

Boerner, Christopher, and Thomas Lambert. 1995. "Environmental Justice Can Be Achieved Through Negotiated Compensation," in *Environmental Justice*, edited by J.S. Petrikin. San Diego: Greenhaven Press (Chapter 7).

Brueggemann, Martin R. 1993. "Environmental racism in our own backyard: Solid waste disposal in Holly Springs, North Carolina." Chapel Hill, N.C.: Master's thesis for the University of North Carolina Graduate School of Journalism.

Bullard, Robert D. 1983. "Solid Waste Sites and the Black Houston Community." *Sociological Inquiry*, (Spring): 273–288.

———. 1992. "Environmental Blackmail in Minority Communities." In *Race and the Incidence of Environmental Hazards: A Time for Discourse*, edited by Bunyan Bryant and Paul Mohai. Boulder, CO: Westview Press (Chapter 6).

———. 1993. "Anatomy of Environmental Racism and the Environmental Justice Movement." In *Confronting Environmental Racism: Voices from the Grassroots*, edited by R.D. Bullard. Boston: South End Press (Chapter 1).

———. 1994a. *Dumping in Dixie: Race, Class, and Environmental Quality*, Second Edition. Boulder, CO: Westview Press.

———. 1994b. *Unequal Protection: Environmental Justice and Communities of Color*. San Francisco: Sierra Club Books.

———. 1994c. *Worst Things First? The Debate over Risk-Based National Environmental Priorities*. Washington, D.C.: Resources for the Future.

Buttel, Frederick R. 1987. "New Directions in Environmental Sociology." *Annual Review of Sociology* 13: 465–488.

Commission for Racial Justice of the United Church of Christ. 1987. *Toxic Wastes and Race in the United States: A National Report on the Racial and Socio-Economic Characteristics of Communities Surrounding Hazardous Waste Sites*. New York: United Church of Christ.

Davis, M. E. 1977. "Occupational Hazards and Black Workers." In *Urban Health* (August): 16–18.

"Dumping on Black America," *Atlanta Constitution*, Editorial, 27 April 1987.

Foster, John Bellamy. 1993. "Let Them Eat Pollution." In *Monthly Review* (January): 10–20.

Freudenberg, N. 1984. "Not In Our Backyards: Community Action for Health and the Environment." *Monthly Review Press* 22:34–39.

Gelobter, Michel. 1992. "Toward a Model of 'Environmental Discrimination.'" In *Race and the Incidence of Environmental Hazards: A Time for Discourse*, edited by Bunyan Bryant and Paul Mohai. Boulder, CO: Westview Press (Chapter 5).

Gianessi, Leonard P., Henry M. Peskin, and Edward Wolff. 1979. "The Distributional Effects of Uniform Air Pollution Policy in the United States." *Journal of Economics* (May):281–301.

Goldman, Benjamin A. 1993. *Not Just Prosperity: Achieving Sustainability with Environmental Justice*. Washington, D.C.: National Wildlife Federation.

Grossman, Mark. 1993. *The ABC-CLIO Companion to the Civil Rights Movement*. Santa Barbara, CA: ABC-CLIO.

Hamilton, James T. 1993. "Politics and Social Cost: Estimating the Impact of Collective Action on Hazardous Waste Facilities." *Fand Journal of Economics* (Spring):101–125.

Harrison, David. 1975. *Who Pays for Clean Air: The Cost and Benefit Distribution of Federal Automobile Emission Controls*. Cambridge, MA: Ballenger Publishing Company.

Hoffman, William S., et al. 1967. "Relation of Pesticide Concentration in Fat to Pathological Changes in Tissues." *Archives of Environmental Health* 15:758–762.

Holtsman, Elizabeth. 1992. *Smokescreen: How the Department of Sanita-*

tion's Solid Waste Plan and Environmental Impact Statement Cover up the Poisonous Health Effects of Burning Garbage. New York: Office of the Comptroller, June. "Jobs and Illness in Petrochemical Corridor." *Washington Post,* 22 December 1989, p. A1.

Johnson, F. Reed. 1980. "Income Distributional Effects of Air Pollution Abatement: A General Equilibrium Approach." *Atlantic Economic Journal* pp. 10–21.

Krauss, Celene. 1994. "Women of Color on the Front Line." In *Unequal Protection: Environmental Justice and Communities of Color,* edited by R.D. Bullard. San Francisco: Sierra Club Books.

LaBalme, J. 1987. *A Struggle for Environmental Justice.* Durham, NC: Regulator Press.

Lavelle, Marianne, and Marcia Coyle. 1993. "Unequal Protection: The Racial Divide in Environmental Law." *The National Law Journal* 21 (September):S1–S12.

Lazarus, Richard J. 1993. "Pursuing 'Environmental Justice': The Distributional Effects of Environmental Protection." *Northwestern University Law Review* 87(3, March 1993):101–170.

Lee, Charles. 1992. "Toxic Waste and Race in the United States." In *Race and the Incidence of Environmental Hazards: A Time for Discourse,* edited by Bunyan Bryant and Paul Mohai. Boulder, CO: Westview Press.

Louisiana Advisory Committee to the U.S. Commission on Civil Rights. 1993. *The Battle for Environmental Justice in Louisiana . . . Government, Industry, and the People.* Kansas City, MO: U.S. Commission on Civil Rights, Central Regional Office.

Lowe, G. D., and T. K. Pinhey. 1982. "Rural-Urban Differences in Support for Environmental Protection." *Rural Sociology* 47:114–128.

Miller, G. Tyler. 1985. *Living in the Environment: An Introduction to Environmental Science, 4th edition.* Belmont, CA: Wadsworth Publishing Company.

Mohai, Paul, and Bunyan Bryant, eds. 1992. "Environmental Racism: Reviewing the Evidence." In *Race and the Incidence of Environmental Hazards: A Time for Discourse.* Boulder, CO: Westview Press (Chapter 13).

Morris, Aldon D. 1984. *The Origins of the Civil Rights Movement: Black Communities Organizing for Change.* New York: The Free Press.

Moses, Marion. 1989. "Pesticide Related Health Problems in Farm Workers." *American Association of Occupational Health Nurses Journal* 37:115–130.

———. 1993. "Farmworkers and Pesticides." In *Confronting Environmental Racism: Voices from the Grassroots,* edited by R.D. Bullard. Boston: South End Press (Chapter 10).

Mpanya, Mutombo. 1992. "The Dumping of Toxic Waste in African Countries: A Case of Poverty and Racism." In *Race and the Incidence of Environmental Hazards: A Time for Discourse,* edited by Bunyan Bryant and Paul Mohai. Boulder, CO: Westview Press (Chapter 15).

Norris, Ruth, ed. 1982. *Pills, Pesticides & Profits: The International Trade in Toxic Substances.* Croton-on-Hudson, NY: North River Press.

Perfecto, Ivette. 1992. "Pesticide Exposure of Farm Workers and the International Connection." In *Race and the Incidence of Environmental Hazards: A Time for Discourse,* edited by Bunyan Bryant and Paul Mohai. Boulder, CO: Westview Press (Chapter 14).

Petulla, Joseph M. 1977. *American Environmental History.* San Francisco: Boyd & Fraser.

Probst, Kate. 1990. *Hazardous Waste Sites and the Rural Poor: A Preliminary Assessment.* Arlington, VA: Clean Sites, Inc.

"The Rise of Anti-Ecology?" *Time,* 3 August 1970, p. 42.

Roberts, Sam. 1992. "In My Backyard? Where New York City Puts Its Problems." *New York Times,* 6 December, p. 54.

Robinson, Ronald. 1994. "West Dallas versus the Lead Smelter." In *Unequal Protection: Environmental Justice and Communities of Color,* edited by R.D. Bullard. San Francisco: Sierra Club Books.

Robinson, Wm. Paul. 1992. "Uranium Production and Its Effects on Navajo Communities Along the Rio Puerco in Western New Mexico." In *Race and the Incidence of Environmental Hazards: A Time for Discourse,* edited by Bunyan Bryant and Paul Mohai. Boulder, CO: Westview Press (Chapter 12).

Skull Valley Goshute Tribe Executive Office. 1993. "Native Americans Have the Right to Make Their Own Land-Use Decisions." In *Environmental Justice,* edited by J.S. Petrikin. San Diego: Greenhaven Press (Chapter 5).

Small, Gail. 1994. "Native Americans Must Fight to Prevent Environmental Injustice on Their Homelands." *The Amicus Journal* (Spring):38–41.

Smith, James Noel, ed. 1974. *Environmental Quality and Social Justice in Urban America.* Washington, D.C.: The Conservation Foundation.

Szasz, Andrew. 1994. *EcoPopulism: Toxic Waste and the Movement for Environmental Justice.* Minneapolis: University of Minnesota Press.

Taylor, Dorceta E. 1989. "Blacks and the Environment: Toward an Explanation of the Concern and Action Gap Between Blacks and Whites." *Environment and Behavior* 21 (March 1989):175–205.

———. 1993. "Environmentalism and the Politics of Inclusion." In *Confronting Environmental Racism: Voices from the Grassroots*, edited by R.D. Bullard. Boston: South End Press (Chapter 3).

Thoreau, Henry David. 1962. "Walking." In *Excursions*. New York: Corinth Books.

U.S. Congress. 1993. *Environmental Justice*. Environmental Justice Hearings before the Subcommittee on Civil and Constitutional Rights of the House Committee on the Judiciary, 103rd Congress, 1st Session, 3 and 4 March.

U.S. Environmental Protection Agency. 1992. *Women, Minorities and People with Disabilities*. Washington, D.C.: Environmental Protection Agency.

U.S. General Accounting Office. 1983. *Siting of Hazardous Waste Landfills and Their Correlation with the Racial and Soci-Economic Status of Surrounding Communities*. Washington, D.C.: U.S. General Accounting Office.

Wasserstrom, R. F., and R. Wiles. 1985. *Field Duty: U.S. Farmworkers and Pesticides Safety. Study 3*. Washington, D.C.: World Resources Institute, Center for Policy Research.

West, Patrick C. 1992. "Invitation to Poison? Detroit Minorities and Toxic Fish Consumption from the Detroit River." In *Race and the Incidence of Environmental Hazards: A Time for Discourse*, edited by Bunyan Bryant and Paul Mohai. Boulder, CO: Westview Press (Chapter 7).

West, Patrick C., et al. 1992. "Minority Anglers and Toxic Fish Consumption: Evidence from a Statewide Survey of Michigan." In *Race and the Incidence of Environmental Hazards: A Time for Discourse*, edited by Bunyan Bryant and Paul Mohai. Boulder, CO: Westview Press (Chapter 8).

Wild, Peter. 1979. *Pioneer Conservationists of Western America*. Missoula, MT: Mountain Press.

Chronology 2

The environmental justice movement is relatively young. Its origin is often traced to the protests held in Warren County, North Carolina, in 1982. Yet, many of the elements of the modern movement can be traced much farther back, at least to the Civil Rights Act of 1875 and the first environmental law of 1899. This chapter summarizes some of the most important events in the history of the modern environmental justice movement.

1863	The Emancipation Proclamation, issued by President Abraham Lincoln, frees all slaves in states in rebellion against the Union.
1875	The Civil Rights Act of 1875 guarantees to all Americans the right of access to public accommodations. For decades, the act is not enforced in most cases involving African Americans and other minorities.
1896	In the case of *Plessy v. Ferguson*, the U.S. Supreme Court rules that

1896
cont.

segregation is not illegal as long as equal accommodations are made available to African Americans.

1899

The nation's first environmental law, the Refuse Act of 1899, is passed. The act was originally written to protect navigable waters from pollution by sediments, disease-carrying organisms, and oil discharges, but it was broad enough to cover many environmental problems through the early 1970s.

1947

The Federal Insecticide, Fungicide, and Rodenticide Act of 1947 is passed. The act originally deals only with pesticides shipped across state lines. The act is amended and extended in 1972 to cover broader aspects of pesticide production and use.

1948

The Water Pollution Control Act of 1948 is passed by Congress. The act is written with good intentions, but is never implemented because almost no funds are appropriated for it.

1956

A stronger version of the 1948 Water Pollution Control Act is adopted. The act is amended and strengthened a number of times in following years, including in 1961, 1965, 1966, and 1970.

1957

The Voting Rights Act of 1957 is passed.

1960

The U.S. Supreme Court rules that segregation in interstate bus and train stations is illegal. A year later, the Interstate Commerce Commission officially prohibits such practices.

1964

The Twenty-fourth Amendment to the U.S. Constitution is adopted. The Amendment abolishes the poll tax, a primary means by which people of color were long prevented from voting, and thus participating in the political process in many southern states.

The Civil Rights Act of 1964 is adopted, guaranteeing to all American citizens equal access to employment and public accommodations, regardless of race, creed, color, sex, or national origin.

1965   The Voting Rights Act of 1965, strengthening its pre-
            decessor, the Voting Rights Act of 1957, is passed.

            The Fair Housing Act of 1968 is adopted. The act is
            later strengthened by amendments adopted in 1988.

1970   The National Environmental Policy Act of 1969 is
            passed by Congress and signed by President Richard
            Nixon. The act is the single most comprehensive leg-
            islative action dealing with environmental issues in
            the nation's history. One of the provisions of the act is
            the creation of the Council on Environmental Quality,
            charged with publishing an annual report on the sta-
            tus of the nation's environmental elements.

            The Clean Air Act of 1970 is passed by Congress and
            signed by the President.

            The U.S. Environmental Protection Agency (EPA) is
            created as an independent agency in the Reorganiza-
            tion Plan No. 3 of 1970.

            Senator Philip Hart (D-Michigan) arranges a meeting
            among environmental groups, labor unions, and mi-
            nority organizations to consider issues of the urban
            environment of concern to all such groups.

            April 22 is declared to be Earth Day. By some esti-
            mates, more than 20 million people demonstrated on
            behalf of improved environmental conditions in the
            United States and the world.

1971   The annual report of the Council on Environmental
            Quality acknowledges that environmental quality is
            affected by the racial and class status of communities.

1972   The Water Pollution Control Act of 1972 is adopted.
            The act represents a giant step forward in the na-
            tion's commitment to protect its water resources
            from pollution.

1975   The Federal Insecticide, Fungicide, and Rodenticide
            Act Amendments of 1975 are adopted.

1976   The Toxic Substances Control Act of 1976 and Re-
        source Conservation and Recovery Act of 1976 are
        adopted.

        A conference on the urban environment is held at
        Black Lakes, Michigan. The conference is organized by
        the United Auto Workers and includes representatives
        from unions, environmental organizations, religious
        groups, and those concerned with economic justice.

        The Urban League creates Project Que: Environmental
        Concerns in the Inner City.

1977   The Clean Air Act and Clean Water Act are revised
        and updated.

        Grants totaling $66,000 permit the Urban Environ-
        ment Conference to fund 11 nationwide conferences
        on issues related to environmental justice.

1979   City Care: A Conference on the Urban Environment is
        held in Detroit, Michigan. Organizers include the Na-
        tional Urban League, the Urban Environment Confer-
        ence, and the Sierra Club. One result of the meeting is
        a sense that traditional environmental groups are
        going to be uncomfortable with the inclusion of social
        justice issues in their agenda.

1982   Residents of Warren County, North Carolina, sup-
        ported by the United Church of Christ, stage a demon-
        stration in opposition to the siting of a polychlorinated
        biphenyl (PCB) landfill near the community of Afton.
        More than 500 African American protestors are ar-
        rested in an unsuccessful attempt to block construction
        of the landfill. This event has been described by some
        authorities as the real beginning of the modern envi-
        ronmental justice movement.

1983   "Taking Back Our Health—An Institute on Surviving
        the Toxic Threat to Minority Communities" is held in
        New Orleans, Louisiana. The conference is sponsored
        by the Urban Environment Conference, shortly before
        it loses its funding and goes out of business.

A report by the General Accounting Office states that three out of four hazardous waste sites in EPA Region 4 are located in African American communities.

1985　　The grassroots committee, People Concerned about MIC [methyl isocyanate], is organized in Institute, West Virginia. Consisting largely of African Americans, the group is created in response to a chemical leak from a nearby Union Carbide plant that resulted in about 135 residents being sent to the hospital.

The EPA commissions the Council of Energy Resource Tribes to conduct a study of potential hazardous waste sites located on or near Indian lands. The survey reveals as many as 1,200 hazardous waste sites on or near 25 Indian reservations.

1987　　The Commission for Racial Justice of the United Church of Christ publishes a report, "Toxic Wastes and Race in the United States," showing that race, even more than income level, is the crucial factor shared by communities exposed to toxic waste.

Four thousand tons of toxic wastes are dumped in Koko, Nigeria. Studies later show that an increase in health problems may be related to this act.

A shipment of 13,476 tons of toxic municipal incinerator ash from Philadelphia is dumped in Haiti. The Haitian government has issued a permit for the import of fertilizer, but the actual delivery consists of toxic wastes.

1988　　The U.S. Supreme Court rules that the development of geothermal energy plants in the Hawaiian Islands does not impinge on the First Amendment right to freedom of religion among native Hawaiians. Native Hawaiians had claimed that the development of such plants would destroy the rain forests and interfere with their worship of the volcanic goddess Pele.

A shipment of garbage and incinerator ash from Philadelphia, originally rejected by both Haiti and

1988
cont.

Panama, is accepted by the Guinean government. Later reports claim that the trees on Kassa Island, where the shipment is dumped, turn brown and die.

The Federal Insecticide, Fungicide and Rodenticide Act is passed, establishing standards for the use of pesticides.

1989

Residents of the "Cancer Alley" section of the lower Mississippi River organize the Great Louisiana Toxic March in an attempt to bring attention to their polluted living conditions.

Thirty-three nations sign the Basel Treaty, an international agreement dealing with the shipment of hazardous materials across international borders. Eventually, more than 100 additional nations become party to the agreement.

Mothers of East Los Angeles (MELA) successfully protest the construction of a $29 million incinerator designed to burn 125,000 pounds of toxic wastes per day.

1990

Students at the Harvard Law School, the New York University Law School, the University of California at Berkeley Law School, and Washington University in St. Louis hold conferences on the issue of environmental justice.

The Michigan Coalition Conference issues a report on "Race and the Incidence of Environmental Hazards."

The EPA establishes the Environmental Equity Workgroup.

A number of grassroots environmental justice groups, including the Gulf Coast Tenants Organization, the Southwest Organizing Project, the United Church of Christ Commission for Racial Justice, and the Southern Organizing Committee for Economic and Social Justice, jointly write to the Big Ten of mainstream environmental groups, expressing their concerns about the groups' historic lack of attention to the special needs and interests of people of color and other minorities and challenging them to end their racist and elitist policies and practices.

New York City adopts a "fair share" act intended to ensure that every part of the city receives a fair share of hazardous facility sitings.

1991   First National People of Color Environmental Leadership Summit is held in Washington, D.C. The Summit adopts a 17-point *Principles of Environmental Justice* statement.

In what is apparently the first significant contribution from the law community on the issue of environmental justice, Rachel D. Godsil, a student, publishes a commentary on the subject in the *University of Michigan Law Review.*

The Agency for Toxic Substances and Disease Registry (ATSDR) initiates a number of actions relating to environmental inequities, including the Minority Environmental Health Conference and a study of minority communities located near National Priorities List hazardous waste landfills.

1992   Students at Columbia University and the Universities of Michigan and Minnesota hold conferences on the issue of environmental justice.

The EPA establishes an Office of Environmental Justice.

The EPA releases a report "Environmental Equity: Reducing Risks for All Communities."

A report in the National Law Journal, "Unequal Environmental Protection," claims that the EPA pursues discriminatory practices in enforcement of environmental laws and regulations.

A national workshop, "Equity in Environmental Health: Research Issues and Needs," is held at Research Triangle Park, North Carolina, under the sponsorship

1992
cont.

of the EPA, ATSDR, and the National Institute of Environmental Health Sciences.

Congressman John Lewis (D-Georgia) and Senator Al Gore (D-Tennessee) introduce the Environmental Justice Act of 1992.

Congresswoman Cardiss Collins (D-Illinois) introduces an amendment to the Resource Conservation and Recovery Act that would require a "community information statement" for the construction of any new hazardous facility. The statement would include a description of the demography of the proposed site as well as a projected estimate of the impact of the facility on the area.

1993

On Earth Day 1993, President Bill Clinton pledges that he will issue an Executive Order instructing federal agencies to take cognizance of the issues raised by the environmental justice movement and to take actions that will reduce the problems emphasized by that movement.

The EPA establishes the National Environmental Justice Advisory Council.

Congressman Lewis and Senator Max Baucus (D-Montana) reintroduce the Environmental Justice Act.

Hearings on the proposed promotion of the EPA to cabinet status include discussion of environmental justice issues.

The Texas Air Control Board and the Texas Water Commission jointly create a statewide Task Force on Environmental Equity and Justice to deal with basic issues, such as reasons that hazardous facilities tend to be located in minority communities, policies and procedures of the two agencies that relate to issues of environmental inequities, methods by which the agencies can become more "user friendly" to communities of color, and data-gathering methods by which the government might become more aware of environmental inequities in hazardous facilities sitings.

1994   President Bill Clinton signs Executive Order 12898 on Environmental Justice ordering federal agencies to abolish and prevent policies that lead to a disproportionate distribution of environmental hazards to communities of color or low income.

An Interagency Symposium on Health Research and Needs to Ensure Environmental Justice, sponsored by the EPA, is held in Arlington, Virginia.

A Federal Interagency Working Group on Environmental Justice is created.

The United Church of Christ issues an update of its 1987 report, "Toxic Waste and Race Revisited," providing further evidence on the relationship between race and toxic waste facilities.

1995   The first public meeting of the Interagency Working Group on Environmental Justice is held in Atlanta, Georgia.

The EPA announces the award of $3 million to assist 174 community-based organizations, tribal governments, and academic institutions to address environmental justice issues in their communities. The awards reflect an increase from the $500,000 awarded to 61 recipients in 1994.

A conference, *Environmental Justice and Transportation: Building Model Partnerships,* is held in Atlanta, Georgia, under the joint sponsorship of the U.S. Department of Transportation and the Clark Atlanta Environmental Justice Resource Center. The conference is a part of the department's public outreach plan as mandated by President Clinton's Executive Order 12898.

Biographical Sketches

The biographical history of the environmental justice movement is quite different in some ways from that of other major movements, such as the traditional environmental and the civil rights movements. The environmental justice movement is largely decentralized, manifested in the activities of dozens or hundreds of local groups, rather than centralized, operating under the direction of large national organizations. As a result, many of the real heroes and heroines of the environmental justice movement are organizers or chairs of town, county, village, or parish organizations; they are people whose names are probably not well known outside of the environmental justice movement itself.

To be fair, this chapter should probably contain biographical sketches of many of those who have led small, local groups as well as of those whose name are more widely known. Since space does not permit the inclusion of all those whose names belong here, the chapter provides a sample of women and men who have made significant contributions to the environmental justice movement, at whatever level they may have worked.

Some important individuals whose biographical sketches have been omitted are listed at the end of this chapter.

Bunyan Bryant (1935–)

An important conference held on the issue of environmental justice was the 1990 Conference on Race and the Incidence of Environmental Hazards. It was sponsored by and held at the University of Michigan. Co-organizer of the conference was Dr. Bunyan Bryant. One consequence of the Michigan conference was a series of high-level policy discussions with U.S. Environmental Protection Agency (EPA) Administrators William K. Reilly and Carol Browner—discussions to which Bryant was an important contributor.

Bryant has also been involved in a number of other seminal meetings and organizations out of which has grown the modern environmental justice movement. For three years, he was cofacilitator of the Martin Luther King Planning Committee at the University of Michigan, where workshops on environmental justice were an integral part of celebrating Dr. King's legacy. Bryant served on the Advisory Committee of the First National People of Color Environmental Leadership Summit in 1991, was cofacilitator of the Symposium for Health Research and Needs to Ensure Environmental Justice in 1994, and was a member of the EPA's National Environmental Justice Advisory Council in 1994–1995.

Bryant was born in Little Rock, Arkansas, on 6 March 1935. He received his B.S. degree in social science from Eastern Michigan University, and his M.S.W. in social work and his Ph.D. in education from the University of Michigan. Bryant did his post-doctoral research in town and country planning at the University of Manchester in England. He is currently on the faculty in the School of Natural Resources and Environment at the University of Michigan as well as serving as a member of the Urban Technological and Environmental Planning Program and the Center of Afro-American and African Studies.

Bryant is author, co-author, and editor of a number of articles and books, including *Environmental Advocacy: Concepts, Issues, and Dilemma; Race and the Incidence of Environmental Hazards: A Time for Discourse* (with Paul Mohai); *Environmental Justice: Issues, Policies, and Solutions;* and *Social and Environmental Change: A Manual for Community Organizing and Action.* In addition to his teaching and writing, Bryant has served as consultant to a number of nonprofit environmental organizations and is in demand as a speaker on the topic of environmental justice.

Robert Bullard

One of the most prolific and articulate writers in the field of environmental justice is Robert Bullard, currently Ware Professor of Sociology and director of the Environmental Justice Resource Center at Clark Atlanta University. Bullard's book *Dumping in Dixie* is regarded by some authorities as one of the most powerful statements on the issue of environmental justice yet to be written. In addition to this work, Bullard has authored or co-authored more than 50 articles and book chapters and has authored, co-authored, or edited eight books. He is also in great demand as a speaker, having presented 30 major addresses in the last five years.

Robert D. Bullard attended Alabama A&M University, from which he received his B.S. in government in 1968. He then received his M.A. in sociology from Atlanta University in 1972 and his Ph.D. in the same field from Iowa State University in 1976. His first work experiences were as an urban planner in Des Moines, Iowa (1971–1974), an administrative assistant at the Office of Minority Affairs, Iowa State University (1974–1975), research coordinator in Polk County, Iowa, and then director of research at the Urban Research Center of Texas Southern University (1976–1978).

In 1976, Bullard was also appointed assistant professor at Texas Southern and, in 1980, was promoted to associate professor. He then went on to hold a series of academic appointments at Rice University (1980), the University of Tennessee (1987–1988), the University of California at Berkeley (1988–1989), the University of California at Riverside (1989–1993), and the University of California at Los Angeles (1993–1994). In 1994, Bullard was appointed to his current position at Clark Atlanta University.

Among the many awards and honors given to Bullard have been the Gustavus Myers Award for the Outstanding Book on Human Rights in the United States (1990); the Environmental Justice Award from the Center for Environment, Commerce, and Energy (1990); the Conservation Achievement Award in Science of the National Wildlife Federation (1990); the Environmental Achievement Award of the CEIP Fund (1990); and the Environmental Justice Award of the Citizens Clearinghouse for Hazardous Wastes (1993).

César Chávez (1927–1993)

By some standards, the senior member of the environmental justice movement may be considered to be César Chávez. Decades before the Warren County protests, the Michigan Conference, or the First National Conference of People of Color Environmental, Chávez was working and organizing to deal with environmental issues (as well as other kinds of problems) faced by farmworkers.

César Estrada Chávez was born on a small farm near Yuma, Arizona, on 31 March 1927. His family lost their farm during the Great Depression, and they joined the exodus to California, looking for a better life. That "better life" turned out to be that of a migrant worker, moving from field to field when and if work was available. Chávez had attended more than 30 different schools before dropping out to continue working full time in the fields. His work there was interrupted briefly when he served in the U.S. Navy during the last two years of World War II.

Chávez's political career began in 1952 when he was invited to become first a volunteer and then a paid staff member with the Community Service Organization (CSO), an organization created by Saul Alinsky to help the poor and politically powerless to develop their own political organizations. Chávez worked his way up the ranks in CSO until becoming general director in 1958. By that time, however, the organization was beginning to change in character, losing contact with its grassroots origins, according to Chávez. Thus, he resigned from CSO in 1962, withdrew his savings from the bank, and created the National Farm Workers Association (NFWA).

The NFWA grew slowly as Chávez organized farmworkers one field at a time, one community after another. It recorded a few small victories in the San Joaquin and Imperial Valley fields of California, winning pay increases, for example, for workers in the Delano area in 1965. The major breakthrough for the NFWA came in 1968, however, when migrant grape pickers in the Delano area went on strike for higher wages. Chávez seized the opportunity to make the strike a national campaign for the workers and for the NFWA itself. He sought the aid and cooperation of major liberal establishments throughout the United States, including civil rights groups, religious organizations, major newspapers and magazines, and political leaders like Robert F. Kennedy. The grape strike continued in one form or another for three decades, winning a number of concessions for workers along the way and establishing NFWA (later renamed the United

Farm Workers of America, AFL-CIO) as the most powerful representative for farmworkers, a position it held for many years.

Among Chávez's specific contributions within the UFWA were the establishment of a Burial Insurance Program (1963); a union newspaper, *El Malcriado*; the Farm Worker Credit Union (1963); a theater group, "El Teatro Campesino" (1965); an artistic group, "El Taller Gráfico" (1965); the National Farm Workers Service Center (1966); the Robert F. Kennedy Farm Worker Medical Fund (1967); the National Farm Workers Health Group (1969); Agbayani Village, a farmworker retirement village (1969); and the Juan de la Cruz Farm Workers Pension Fund (1973).

César Chávez died in San Luis, Arizona, on 23 April 1993.

Benjamin F. Chavis (1948–)

A name that stands out in much of the early literature of the environmental justice movement is that of Benjamin F. Chavis. Chavis was one of the "outsiders" invited to participate in the Warren County protests of 1982 against the siting of a PCB (polychlorinated biphenyl) dump. At the time, he was executive director of the Commission for Racial Justice of the United Church of Christ. After his participation in the Warren County protest, he reported that he began to get calls from "all over the country . . . from people telling me that the situation in Warren County was not unusual."

In response to such calls, Chavis commissioned a study under the auspices of the Commission for Racial Justice to determine the relationship between hazardous waste sites and low-income communities and communities of color. That study, released in 1987, became one of the most significant events in the early history of the environmental justice movement. Over the next decade, Chavis continued to act as one of the primary spokespersons on the issue of environmental justice. During his appearance before the House Subcommittee on Civil and Constitutional Rights of the Judiciary Committee on 3 March 1993, he suggested a definition for the term *environmental racism* that is now widely used and quoted (see chapter 8).

Benjamin Franklin Chavis, Jr., was born in Oxford, North Carolina, on 22 January 1948. He attended St. Augustine's College in Raleigh and then the University of North Carolina at Charlotte, from which he received a B.A. degree in chemistry in 1969. He took a job teaching high school chemistry in Oxford, but quickly became bored with the job. He was more interested, he later explained, "in talking civil rights to my students." Thus, he left teaching to take a position as head of the Washington, D.C., office of the Commission for Racial Justice.

In 1976, Chavis earned his master of divinity degree from Duke University under somewhat unusual circumstances. At the time, he was serving a prison sentence for conviction on conspiracy and arson charges brought about during a civil rights battle in Wilmington, North Carolina, five years earlier. He was one of the Wilmington Ten accused of burning down a grocery store during a period of severe racial tensions in Wilmington.

When three important prosecution witnesses announced in 1979 that they had lied during the Wilmington Ten case, Chavis and his colleagues were released. He then traveled throughout the country attempting to aid other wrongfully incarcerated prisoners. This period of his life came to an end in 1986 when he was appointed to his post with the Commission for Racial Justice. Chavis held this position until 1993 when he was appointed to replace Benjamin L. Hook as executive director of the National Association for the Advancement of Colored People. He remained in this position until 1994 when he was replaced by a vote of the board of directors.

Jay Feldman (1953–)

A prominent spokesman for issues involving farmworker health for more than two decades is Jay Feldman. Feldman was one of the founders of the National Coalition Against the Misuse of Pesticides (NCAMP) in 1981 and has served as the organization's executive director since that time. In this role, he provides policy direction and overall coordination of NCAMP's activities aimed at controlling the use of pesticides and promoting nonchemical alternatives in agriculture. He is also writer and editor for NCAMP's newsletter, "Pesticides and You."

Jay Feldman was born in Brooklyn, New York, on 9 May 1953. He earned his B.A. degree in political science from Grinnell College in 1975 and his M.A. in urban and regional planning from Virginia Polytechnic Institute in 1977. In 1974, he was researcher for Associated Colleges of the Midwest in San Jose, Costa Rica. From 1977 to 1979, Feldman worked as a rural health specialist at Rural America, a national advocacy organization for people living in small towns and rural areas.

Among Feldman's many books and other publications are *Rural Health Directory: A Resource Guide to Nongovernmental Organizations Involved in Rural Health* (1978), *Health Platform for Rural*

America (1977), *Pesticides and You* (1980), *Pesticide Use and Misuse: Farmworkers and Small Farmers Speak on the Problem* (1980), and many NCAMP publications, such as *A Failure to Protect, Safety at Home*, and *Unnecessary Risks*.

Feldman has written articles on pesticide use for a number of newspapers and journals, including the *New York Times, Washington Post, Los Angeles Times, Journal of Arboriculture*, and *Environmental Law Reporter*. He has also appeared on both local and national television programs, such as *Nightline, McNeil-Lehrer News Hour, Good Morning America, Today Show*, and *Cable News Network*.

Deeohn Ferris (1953–)

For nearly two decades, Deeohn Ferris has worked on issues of environmental law. During the last quarter of that period, she has focused on the topic of environmental justice.

Deeohn Ferris was born on 3 November 1953 in Norwalk, Ohio. She received her B.A. from Ashland University in 1975 and her J.D. from Georgetown University Law Center in 1978. Her first job was with the U.S. Environmental Protection Agency (EPA) where she worked as attorney-advisor in the Office of Legislation (1979–1983), as assistant enforcement counsel in the Office of Enforcement and Compliance Monitoring (1983–1984), and as Director of the Special Litigation Division (1984–1986). Her key accomplishments at EPA included civil litigation aimed at fostering compliance with chemical assessment and control programs, implementing the first full-scale environmental and management audits, and demonstrating effective new federal compliance tools.

From 1986 to 1989 Ferris worked as environmental counsel on liability and toxic tort issues at the American Insurance Association. She then moved to the National Wildlife Federation (NWF) as director of Environmental Quality from 1990 to 1992. At NWF, she published a quarterly newsletter, "The Gene Exchange," co-authored a national report, *Waters at Risk: Keeping Clean Waters Clean*, and was awarded the Charlie Shaw Award for exceptional contribution to the shared mission of NWF and its affiliates.

In 1992, Ferris accepted a position as program director of the Environmental Justice Project at the Lawyer's Committee for Civil Rights Under Law. While there she organized and launched a national environmental justice project that provides legal and technical assistance to clients and legislative and regulatory advocacy. During this period she was also appointed to the EPA Federal Environmental Justice Advisory Committee. In 1993, Ferris conceptualized the Executive Order on Environmental Justice issued by President Clinton in 1994.

In 1994, Ferris was appointed director of the newly created Alliance for the Washington Office on Environmental Justice, an international collaboration of community-based regional networks and organizations. She is a popular speaker on law, public health, and policy, and author of more than a dozen articles and book chapters on environmental law issues.

Neftalí García Martínez (1943–)

Neftalí García Martínez is the founder and director of Scientific and Technical Services, Inc., (Servicios Cientificos y Tecnicos) in Hato Rey, Puerto Rico. He holds a Ph.D. in chemistry, with a secondary concentration in biochemistry. Martínez has 20 years of experience as a scientific, environmental, economic, and political advisor in a wide variety of areas, including water, soil, and air pollution; hazardous chemicals; evaluation and preparation of environmental impact documents; environmental and occupational health; environmental planning, research, and education; and alternative economic and social projects.

His teaching experience includes assignments at the University of Puerto Rico, the Caguas City College in Caguas, Puerto Rico, and the Rio Piedras Campus of the University of Puerto Rico. He has also served visiting professorships at Queens College of the City University of New York and the State University of New York at Binghamton. In addition to chemistry and biochemistry, Martínez has taught environmental conflicts and the economic history of Puerto Rico.

Included among Martínez's reports and publications are *Analysis of the Economic, Environmental and Health Aspects of the Vegetable Project in Santa Isabel* (1985), "Comments to the Preliminary Environmental Impact Statement of the Caribe Associates Residential Project in Loiza (1985)," "Statement about the Incineration Project of the Trofe Company in Peñuelas (1987)," and *The Export by Industrialized Countries of Unacceptable Hazardous Products to Non-Developing [sic] Countries.*

Martínez was born in Trujillo Alto, Puerto Rico, on 5 May 1943.

Clarice E. Gaylord (1943–)

First director of the U.S. Environmental Protection Agency's (EPA) Office of Environmental Justice is Dr. Clarice E. Gaylord. Gaylord has served with the EPA since 1984 when she became director of the Research Grants staff of the EPA. In 1988, Gaylord was selected for the EPA's Senior Executive Training Program and was then appointed Deputy Director of the Office of Human Resources Management at EPA. When the Office of Environmental Justice was established in 1992, Gaylord was appointed its first director.

Gaylord was born on 14 April 1943. She was awarded her B.A. in zoology from the University of California at Los Angeles in 1965 and her M.S. and Ph.D., both in zoology, from Howard University in 1967 and 1971, respectively. She was research assistant in biochemical genetics and instructor in zoology at Howard from 1968 to 1969. In 1976 Gaylord took a position as geneticist at the National Cancer Institute (NCI). Two years later, she became Health Scientist administrator and then administrator of the Division of Research Grants at NCI.

Among the honors awarded to Gaylord are the EPA Gold and Silver Medals (1992 and 1994), the NIH Award of Merit (1984), the EPA Special Achievement Award (1987), and the Howard University Distinguished Alumni of the Year Award (1989). She is a member of Sigma Xi Research Society, the Beta Kappa Chi Scientific Honor Society, and Sigma Delta Epsilon Graduate Women's Scientific Fraternity.

Juana Beatriz Gutierrez (1932–)

Mother of nine children, grandmother of ten, and community activist *par excellance*: that is one description of Juana Beatriz Gutierrez, founder and current president of Madres del Este de Los Angeles (Mothers of East Los Angeles; MELA). Gutierrez immigrated to the United States from her native Zacatecas, Mexico, in 1952. Almost immediately, she became active in community organizations and programs, explaining that she views the children of her community as part of her own extended family.

In 1985, Gutierrez was asked by California State Assemblywoman Gloria Molina to organize opposition to the proposed construction of a new prison in East Los Angeles. At the time, the community already housed three-quarters of the State's penal population, and Gutierrez, Molina, and their friends refused to accept the notion that East Los Angeles should continue to be the State's dumping grounds for penal institutions. MELA rapidly grew to become the largest single community mobilization since the days of anti-Vietnam War protests.

As president of MELA, Gutierrez has been active in a number of other community activities, including the Whittler Boulevard Rehabilitation Program to beautify the business district, the Neighborhood Watch programs for Dolores Mission and Grade Vista area residents, the Coalition Against Incinerator in Vernon, Jobs with Peace campaign against Proposition V, the Community Against the Lancer Project (a proposed toxic waste incinerator), Coalition Against the Pipeline, California Communities Against Toxins, Community and Labor Strategy Watchdog, and the Kettleman City fight against a toxic incinerator.

Gutierrez's work has been recognized by a number of honors and awards including her selection as Democrat of the Year in the 56th Assembly District; Woman of the Year for 1988, as chosen by the California State Legislature; the Boyle Heights Sports Center Appreciation Award for 1989; the Pope John Paul II Benemereti Award for 1992; and the Mexican Mother of the Year for 1993.

LaDonna Harris (1931–)

LaDonna Harris' biographical profile introduces her as "a remarkable statesman and national leader who has enriched the lives of thousands. She has devoted her life to building coalitions that create change. She has been a consistent and ardent advocate on behalf of Tribal America. In addition, she is active in the civil rights, environmental, women's, and world peace movements."

LaDonna Harris was born on a small farm outside of Walters, Oklahoma, in 1931. She was raised by her maternal grandparents and spoke only Comanche when she entered grade school. She has no formal education beyond high school, from which she graduated in 1949, but she has received seven honorary degrees from colleges and universities such as the University of Oklahoma, Northern Michigan University, Antioch College, Dartmouth College, and Marymount College of New York.

Harris began her career in public service as the wife of U.S. Senator Fred Harris. In the past three decades, she has accumulated an impressive list of accomplishments, including the founding of Oklahomans for Indian Opportunity in 1965, founding of the Council for Energy Resource Tribes in 1976, and founding of Americans for Indian Opportunity in 1970. She was also involved in the establishment of the National Women's Political Caucus, the National Indian Housing Council, the National Indian Business Association, and the National Tribal Council on the Environment. Harris has been guest lecturer at the Aspen Institute, the American Program Institute, the Chautauqua Summer Series on Racism, and the Washington School of the Institute for Policy Studies.

The list of awards given to LaDonna Harris has been extensive, including the Outstanding American Citizen Award, the United Nations Peace Medal, the Ladies' Home Journal Woman of the Year Award, the Human Rights Award of Delta Sigma Theta Sorority and a similar award from the National Education Association, the Mary Church Terrell Award for Distinguished Public Service of Delta Sigma Theta Sorority, and the Award for Public Service of Theta Sigma Phi, National Journalism Fraternity for Women. Harris sits on more than a dozen boards, including those of Common Cause, the Jacobson Foundation, the Keystone Center, the National Urban League, the White House Fellows Advisory Board, and the National Organization of Women.

Harris' more than two dozen publications include *Partnerships for the Protection of Tribal Environments* (1991), *Tribal Governments in the U.S. Federal System* (1990), *A Resource Bibliography for Tribal Participation in Environmental Protection Activities* (1990), *Survey of American Indian Environmental Protection Needs on Reservation Lands* (1986), and *To Govern and To Be Governed: American Indian Tribal Governments at the Crossroads* (1983).

Pamela Tau Lee

Lee has been labor coordinator at the Labor Occupational Health Program (LOHP) in the School of Public Health at the University of California at Berkeley since 1990. In the position, she is responsible for training, technical assistance, materials development, and the planning of conferences on occupational health issues. She also coordinates outreach efforts to labor, community, and environmental justice groups.

Prior to taking her current job at LOHP, Lee was staff director of the Hotel Employees/ Restaurant Employees Union Local 2 AFL-CIO in San Francisco (1985–1990). She received her B.S. degree in sociology from California State University at Hayward in

1970 and attended the Secondary Education Teacher Post Graduate Program at the California State University at Hayward Teacher Corps in 1972.

Lee serves on the boards of directors of the Washington Office on Environmental Justice and the Asian Pacific Environmental Network, and is or has been a member of the Clinton Administration EPA Transition Team, the National People of Color Environmental Summit, and the EPA National Environmental Justice Advisory Council. She has been awarded a Tides Foundation Fellowship for sabbatical research on environmental justice and leadership development in the environmental movement and the American Public Health Association Lorin E. Kerr Occupational Section Award.

Les Leopold (1947–)

For more than a decade, Les Leopold has been actively involved in research on, teaching of, and curriculum development in the field of courses offered to working adults, especially in the field of economics, health, and safety in the workplace. He has also been co-producer and co-author of a number of institute films, booklets, and slide shows on these topics.

Les Leopold was born on 25 May 1947 in Vineland, New Jersey. He received his B.A. degree from Oberlin College in 1969 and his M .A. degree from the Woodrow Wilson School of Public and International Affairs at Princeton University in 1975. His first professional assignment was as assistant dean of students at Pitzer College in Claremont, California, from 1971 to 1972. He was then appointed instructor and special assistant to the president at Pitzer from 1972 to 1973. During the summer of 1974, Leopold worked as project director at the Citizenship Legislative Department of the Oil, Chemical, and Atomic Workers in Washington, D.C.

In 1975, Leopold accepted a position as vice-president of the Labor Institute. Nine years later he became president of the Institute. Among the publications and audiovisual programs for which he is responsible or partially responsible are "Trust in Training," "Your Job or Your Life," "Should Energy Cost an Arm and a Leg?," "Bhopal, USA?," and "Working People and Taxes in New Jersey" (all videotape or slide shows) and *Toxic Jobs and Environment: Perspectives on New Jersey's Toxic Economy, OCAW-Labor Institute Hazardous Waste Training Handbook,* and *Jobs and the Environment Workbook.*

John Lewis (1940–)

"Since my earliest days as an Atlanta city councilmember, I have worked for clean, green spaces in our cities and cleanup of toxic waste sites. I believe this earth is not ours to waste and hoard; we must use our limited resources carefully and leave this planet a little cleaner and greener for future generations." In those words, Congressman John Lewis has described his long-term commitment to the nation's environment. One of the most important manifestations of that commitment was the Environmental Justice Act that Lewis co-sponsored with then Senator Albert Gore in 1992. The bill was an attempt to deal with the cumulative effects of air, water, and soil pollution in American communities. It also called for health assessments in the 100 most polluted areas of the country.

John Lewis was born the son of sharecroppers on 21 February 1940 outside Troy, Alabama. He attended segregated public schools in Pike County, Alabama, while continuing to work on the family farm. He attended Fisk University, from which he received a B.A. degree in religion and philosophy, and then was graduated from the American Baptist Theological Seminary in Nashville, Tennessee. While a student in Nashville, Lewis organized sit-in demonstrations at segregated lunch counters. He later became one of the founders of the Student Nonviolent Coordinating Committee.

During the 1960s, Lewis was actively involved in a number of civil rights activities, including the Freedom Rides designed to end segregation at bus terminals and the Mississippi Freedom Summer.

In 1977, Lewis accepted an appointment by President Jimmy Carter to become director of ACTION, the federal volunteer agency. Three years later, he resigned his post at ACTION to become community affairs director of the National Consumer Co-op Bank of Atlanta. From 1981 to 1986, Lewis served on the Atlanta City Council. He resigned from that post in 1986 to run for Congress from Georgia's Fifth Congressional District, a position he has held since that time.

Lewis currently serves on the House Ways and Means Committee, is co-chair of the Congressional Urban Caucus, and is chief deputy minority whip for the Democratic party.

Paul Mohai (1949–)

One of the most active researchers in the field of environmental inequities has been Paul Mohai. For more than a decade, Mohai has analyzed the social, political, and organizational processes that influence and shape environmental and natural resource policy. Most recently his work has focused on studying public attitudes toward environmental and natural resource issues, analyzing the factors affecting political activism, assessing the impact of environmental pollution and resource scarcities on low-income and minority groups, and evaluating the contribution of wildland recreation to the quality of life.

Paul Mohai was born on 20 June 1949 in Linz, Austria. For his undergraduate education, he attended Michigan State University (1967–1968), the University of Birmingham, England (1969–1970), and the University of California at Berkeley (1968–1971), from which he received his B.A. in mathematics. He then earned his M.S. degree in forestry and statistics from Syracuse University in 1976 and his Ph.D. in natural resource policy and the sociology of natural resources from Pennsylvania State University in 1983.

Mohai's first professional appointment was as mathematics teacher at Fox Lane Middle School in Bedford, New York. He then held assignments as teaching assistant and instructor at the State University of New York College of Environmental Science and Forestry at Syracuse (1973–1976), as project intern and program assistant at the Natural Resources and Forestry Group of the U.S. Department of Agriculture's Cooperate State Research Service (1977–1979), as instructor in the School of Forest Resources at Penn State (1979–1983), and as assistant professor in the College of Natural Resources, Utah State University (1983–1987).

In 1987, Mohai accepted an appointment in the School of Natural Resources and Environment at the University of Michigan. He now holds the title of associate professor at Michigan. He also serves as graduate admissions coordinator in the Policy Program at the School of Natural Resources and Environment. During 1995, Mohai was visiting professor in the Department of Rural Sociology at the University of Wisconsin, Madison.

Mohai has presented more than three dozen invited addresses on environmental justice and other issues since 1983, has conducted nearly two dozen funded research projects, and authored, co-authored, or edited more than 60 books, papers, articles, and reports. Perhaps the most important of these is *Race and the Incidence of Environmental Hazards: A Time for Discourse*, which Mohai co-edited with his colleague at Michigan, Bunyan Bryant.

Richard Moore

Moore has been involved in a variety of social issues for more than 25 years. He has worked with community-based groups around issues such as welfare rights, police brutality, street gang activities, drug abuse, low-cost health care, child nutrition, and the fight against racism. He has been active in the environmental justice movement from its earliest days, serving on the planning committee for the First National People of Color Environmental Leadership Summit in October 1991. He currently serves on the board of directors of the Environmental Support Center, is chair of the National Environmental Justice Advisory Committee, and is a member of the Eco-Justice Working Group of the National Council of Churches.

Moore was the founder of the Bobby Garcia Memorial Clinic in Albuquerque, New Mexico, and a founding member of the Southwest Organizing Project, a multi-racial organization working to empower the disenfranchised in the southwest, and a founding member of the Southwest Network for Environmental and Economic Justice, a coalition of community-based grassroots organizations in six southwestern states and Mexico.

Marion Moses

Founder and currently director of the Pesticide Education Center in San Francisco, California, is Marion Moses, M.D. Moses was born on 24 January 1936 in Wheeling, West Virginia. She was educated at the Georgetown University School of Nursing, from which she earned her B.S.N. in 1957, and at the Columbia University Teachers College, from which she earned an M.A. in education in 1960.

Moses says that she first became interested in the problems of farmworkers in 1964 when she was a student at the University of California at Berkeley. Pictures of the difficult living and working conditions of migrant workers inspired her to begin volunteer work with the Citizens for Farm Labor, a group working to increase public awareness of the plight of migrant workers. During this period, she was involved in reactivating the Student Committee on Agricultural Labor at Berkeley.

As a result of her experiences at Berkeley, Moses decided to volunteer as a nurse with the United Farm Workers (UFW) of America in Delano, California. During the five years she remained with the UFW, Moses was very involved in the union's grape strike, which continued through much of this period. While serving with the UFW, Moses also worked part-time as a research and administrative assistant at the Kaiser Foundation Research Institute in San Francisco.

After finishing her work with the UFW, Moses returned to school and worked toward a medical degree. She entered the Temple University School of Medicine, from which she received her M.D. in 1976. She then completed her residency in internal medicine at the University of Colorado Medical Center in Denver in 1977 and her residency in occupational medicine at the Mount Sinai Medical Center in New York City in 1980. She received her board certification from the American College of Preventive Medicine and Public Health in 1980.

After completing her medical training, Moses returned to the UFW, where she was appointed medical director of the National Farm Workers Health Group in Keene, California. She held that post from 1983 to 1986. She simultaneously held an appointment as assistant clinical professor in the Department of Family and Community Medicine at the University of California School of Medicine in San Francisco, a post she held until 1992. Between 1989 and 1994, Moses was also an adjunct faculty member at the San Diego State University Graduate School of Public Health in the Department of Occupational Medicine. In 1988, she founded and became president of the Pesticide Education Center.

Moses has been appointed to a number of important committees in the field of occupational health, including the administrator's Toxic Substances Advisory Committee of the U.S. Environmental Protection Agency (EPA), the National Advisory Committee of Health Related Effects of Herbicides of the Veterans Administration, the National Advisory Committee of the Pesticide Farm Safety Center of the EPA, the National Advisory Committee of the First National People of Color Environmental Leadership Summit, and as chair of the Pesticide Exposure Workshop of the Environmental Justice: Research Issues and Needs Conference in Arlington, Virginia. Moses is the author of a number of articles, book chapters, and books on occupational health and pesticide exposure.

Peggy Shepard

Peggy Shepard has had two careers, one as a journalist and one as a political activist. She began her first career in September 1969, as a reporter and home furnishings editor at the *Indianapolis News*. From there, she moved on to book publishing, as text and photo researcher for *Time-Life Books* (1971–1973), copy editor for the *San*

Juan Star (1974–1975), associate editor of *Redbook* (1976–1977), and editorial director of *Verve* (1977–1978).

In 1978 Shepard began a long career of government-related work when she accepted a position as consumer affairs specialist at the New York State Division of Housing and Community Renewal. She later became consumer affairs director at the division (1980–1983), special assistant to the commissioner (1983–1985), director of public information for rent administration (1985–1988), and special assistant for Government and Community Affairs (1988–1993). She was then appointed women's coordinator for the New York City Office of the Comptroller.

Shepard's life took a dramatic turn in 1988 when she co-founded with Vernice Miller the West Harlem Environmental Action (WHE-ACT). WHE-ACT was created during the effort to prevent the construction of the North River sewage treatment plant and a Metro Transit Authority bus stop in northern Manhattan. The organization was the first African American and Hispanic American community-based group for educating and mobilizing residents about environmental health and quality-of-life issues exclusively. She has served as executive director of WHE-ACT since November 1994.

In addition to her work at WHE-ACT, Shepard has been active in other political roles. Since 1992 she has served as president of the Manhattan chapter of the National Women's Political Caucus. She is also co-founder of the West Harlem Independent Democrats (1985) and the New York Women of Color for Political Empowerment and was Democratic district leader from 1985 to 1993. Among her awards are the 1970 Indiana Press Club Award, the 1971 Burlington House Award, the 1990 New Yorker for New York Award from Citizens Committee of New York, the 1991 Life of the City Award from *New York Woman* magazine, the 1993 Environmental Women of Action Award from Tambrands, Inc., the 1993 Earthling Award for Environmental Justice of the City Club of New York, and the 1993 People Who Make a Difference Award from *National Wildlife* magazine.

Shepard received her B.A. degree in English from Howard University and has taken courses in management at the Baruch College.

Cora Tucker (1940–)

Symbolic of the spirit and practice of the environmental justice movement is Cora Tucker, founder of Citizens for a Better America.

Born on 12 December 1940 in Halifax County, Virginia, Tucker left school before graduating in order to get married. Fifteen years later, with her seven children all in school, Tucker earned her high school diploma.

At about the same time, Tucker joined with a number of black students in Halifax County to form Citizens for a Better America (CBA), an organization whose goal it is "to make America a better place for all." Building on its initial effort to obtain recreational facilities for black students in the county, CBA expanded its efforts to ensure fair and equal treatment for people of color by pushing for road paving, integrating the Agricultural Extension Agency's Home Demonstration Clubs, organizing voter registration drives, conducting surveys of public and private minority employment practices, serving formal complaints against unfair hiring practices, and monitoring local tax policies.

A scrapbook of articles about Tucker's life clearly show that her name is anathema to many public officials, but she has obviously earned a place in the heart of many residents of Halifax County. (The scrapbook is available from the Citizens for a Better America, address in chapter 5.)

Beverly Wright (1947–)

A member of the Michigan Coalition, the National Advisory Committee for the First National People of Color Environmental Leadership Summit, the Planning and Protocol Committees for the National Institute for Environmental Health Sciences' Health and Research Needs to Ensure Environmental Justice National Symposium, the EPA's National Environmental Justice Advisory Council, and founder and director of the Deep South Center for Environmental Justice at Xavier University: these are some of the major activities in which Beverly Hendrix Wright has been engaged over the past 15 years. She has also been active in other professional activities sponsored by the National Council for Negro Women, the Association of Black Psychologists, the Association of Social and Behavioral Scientists, and the Society for the Study of Social Problems.

Wright was born in New Orleans on 1 October 1947. She earned her B.A. degree in sociology from Grambling College in 1969 and her M.A. (1971) and Ph.D. (1977), both in sociology, from the State University of New York at Buffalo. She served as instructor in the Departments of Sociology at Millard Fillmore College (1970–1971) and the State University of New York at Buffalo

(1970–1974). She was appointed to the faculty at the University of New Orleans in the Department of Sociology in 1974 and then was promoted to assistant professor in 1977 and associate professor in 1987. During the period from 1989 to 1993 she served as associate professor of sociology at Wake Forest University. In 1993, she was appointed director of the Deep South Center for Environmental Justice at Xavier University of Louisiana in New Orleans.

Wright is the author or co-author of two dozen papers, reviews, and book chapters. She has also presented about three dozen papers at professional meetings. Among her honors and awards are selection for Who's Who of Outstanding Young Women in America (1979), Urban Research Fellow at Howard University's Institute for Urban Affairs and Research (1979), Urban League Outstanding Achievement Award (1987), and Environmental Justice Award from the Tulane University Chapter of the Black Law Students Association and from the Citizens Clearinghouse for Hazardous Waste (both in 1993).

Concluding Note

The names of some individuals who have made important contributions to the environmental justice movement have been omitted from this chapter because they did not respond to numerous requests for information from which a biographical sketch could be written. Their omission from this chapter is by no means a reflection of their role in the environmental justice movement, but only a consequence of the author's having insufficient information from which to write a complete sketch. The following individuals fall into this category:

Pat Bryant, Gulf Coast Tenants Organization
Tom Goldtooth, Red Lake Band of Chippewa
Hazel Johnson, People for Community Recovery
Charles Lee, Commission for Racial Justice of the United Church of Christ
Vernice Miller, formerly West Harlem Environmental Action; now, Natural Resources Defense Council
Peggy Saika, Asian Pacific Environmental Network
Gail Small, Native Action
Connie Tucker, Southern Organizing Committee for Economic and Social Justice

Documents 4

Legislative and legal issues are an important part of the environmental justice movement. To a large extent, obtaining redress for environmental inequities and preventing such inequities from occurring in the future mean that people must know about existing environmental and civil rights laws and know how to use those laws. It also means having new laws written that will accomplish these same goals.

This chapter includes a number of documents important to this aspect of the environmental justice movement. In the first section of the chapter, some important laws, treaties, bills, and executive orders bearing on issues of environmental justice are presented. The second section of the chapter includes some court cases whose decisions have had bearing on environmental inequities. Finally, in the third section of the chapter, a few of the best-known recommendations dealing with environmental justice have been reprinted.

Laws, Treaties, Bills, and Executive Orders

In his executive order on environmental justice of 11 February 1994, President Bill Clinton pointed out that a host of federal laws was already available for those who believed that they were the victim of environmental injustice. As the name of the movement itself suggests, those laws fall into two large categories, those dealing with civil rights and those dealing with environmental protection. This section contains a number of laws and executive orders on which proponents of environmental justice have based their appeals for action and, in many cases, court cases brought in the name of environmental injustice. The section concludes with the texts of one of the bills that have been submitted to the U.S. Congress dealing specifically with the issue of environmental justice.

The Civil Rights Act of 1964

The fundamental statement of the nation's position on civil rights is contained in the Civil Rights Act of 1964. This sweeping piece of legislation covered nearly every aspect of American public life including institutionalized persons (Chapter I), public accommodations (Chapter II), public facilities (Chapter III), education (Chapter IV), and employment (Chapter VI). The first Subchapter of this law lays out clearly the philosophy on which the law is based. It reads as follows:

Subchapter I—Generally

1981. Equal rights under the law

All persons within the jurisdiction of the United States shall have the same right in every State and Territory to make and enforce contracts, to sue, be parties, give evidence, and to the full and equal benefit of all laws and proceedings for the security of persons and property as is enjoyed by white citizens, and shall be subject to like punishment, pains, penalties, taxes, licenses, and exactions of very kind, and to no other.

1982. Property rights of citizens

All citizens of the United States shall have the same right, in every State and Territory, as is enjoyed by white citizens thereof to inherit, purchase, lease, sell, hold, and convey real and personal property.

This portion of the Civil Rights Act was codified in Executive order No. 11063, "Equal Opportunity in Housing." The critical section of that order reads as follows:

Part I—Prevention of Discrimination

Section 101. I hereby direct all departments and agencies in the executive branch of the Federal Government, insofar as their functions related to the provision, rehabilitation, or operation of housing and related facilities, to take all action necessary and appropriate to prevent discrimination because of race, color, religion (creed), sex or national origin—

(a) in the sale, leasing, rental, or other disposition of residential property and related facilities (including land to be developed for residential use), or in the use or occupancy thereof, if such property and related facilities are—

(i) owned or operated by the Federal Government, or

(ii) provided in whole or in part with the aid of loans, advances, grants, or contributions hereafter agreed to be made by the Federal Government, or

(iii) provided in whole or in part by loans hereafter insured, guaranteed, or otherwise secured by the credit of the Federal Government, or

(iv) provided by the development or the redevelopment of real property purchased, leased, or otherwise obtained from a State or local public agency receiving Federal financial assistance for slum clearance or urban renewal with respect to such real property under a loan of grant contract hereafter entered into; and

(b) in the lending practices with respect to residential property and related facilities (including land to be developed for residential use) of lending institutions, insofar as such practices relate to loans hereafter insured or guaranteed by the Federal Government.

The so-called "Title VI" provision of the act (actually Subchapter V in the law) has been the piece of legislation most commonly called upon by those working for environmental justice. Title VI bans discrimination in federally funded programs and has been used by residents of environmentally disadvantaged communities to prosecute industries that pollute their communities. The following passages provide a flavor of the guarantees of civil liberties found in this act.

Subchapter V—Federally Assisted Programs

2000d. Prohibition against exclusion from participation in, denial of benefits of, and discrimination under federally assisted programs on ground of race, color, or national origin.

No person in the United States shall, on the ground of race, color, or national origin, be excluded from participation in, be denied the benefits of, or be subjected to discrimination under any program or activity receiving Federal financial assistance.

2000d-1. Federal authority and financial assistance to programs or activities by way of grant, loan, or contract other than contract of insurance or guaranty; rules and regulations; approval by President; compliance with requirements; reports to Congressional committees; effect date of administrative action.

Each Federal department and agency which is empowered to extend Federal financial assistance to any program or activity, by way of grant, loan, or contract other than a contract of insurance or guaranty, it is authorized and directed to effectuate the provisions of section 2000d of this title with respect to such program of activity by issuing rules, regulations, or orders of general applicability which shall be consistent with achievement of the objectives of the statute authorizing the financial assistance in connection with which the action is taken. No such rule, regulation, or order shall become effective unless and until approved by the President. Compliance with any requirement adopted pursuant to this section may be effected (1) by the termination of or refusal to grant or to continue assistance under such program or activity to any recipient as to whom there has been an express finding on the record, after opportunity for hearing, of a failure to comply with such requirements, but such termination or refusal shall be limited to the particular political entity, or part thereof, or other recipient as to whom such a finding has been made and, shall be limited in its effect to the particular program, or part thereof, in which such noncompliance has been so found, or (2) by any other means authorized by law: *Provided, however,* that no such action shall be taken until the department or agency concerned has advised the appropriate person or persons of the failure to comply with the requirement and has determined that compliance cannot be secured by voluntary means. In the case of any action terminating, or refusing to grant or continue, assistance because of failure to comply with a requirement imposed pursuant to this section, the head of the Federal department or agency shall file with the committees of the House and Senate having legislative jurisdiction over the program or activity involved a full written report of the circumstances and the grounds for such action. No such action shall become effective until thirty days have elapsed after the filing of such report.

Source: U.S.C. Title 42, Chapter 21, p. 372 et seq., 1981 et seq.

The Fair Housing Act of 1968

Some experts working in the field of environmental justice believe that the Fair Housing Act of 1968 will provide a strong justification for cases brought against companies responsible for the pollution of neighborhoods and communities. They argue that the act calls for all residents to receive equal treatment in the United States, but that the presence of disproportionate siting of polluting and hazardous waste sites violates that principle. Section 3601 of the act lays out the general philosophy behind the act, and then section 3604 outlines some of the details of nondiscrimination required by the act. Probably the most important part of the following selection is section 3604(b).

3601. Declaration of policy
It is the policy of the United States to provide, within constitutional limitations, for fair housing throughout the United States. . . .

3604. Discrimination in the sale or rental of housing and other prohibited practices
As made applicable by section 3603 of this title and except as exempted by sections 3603(b) and 3607 of this title, it shall be unlawful—

(a) To refuse to sell or rent after the making of a bona fide offer, or to refuse to negotiate for the sale or rental of, or otherwise make unavailable or deny, a dwelling to any person because of race, color, religion, sex, familial status, or national origin.

(b) To discriminate against any person in the terms, conditions, or privileges of sale or rental of a dwelling, or in the provision of services or facilities in connection therewith, because of race, color, religion, sex, familial status, or national origin.

(c) to make, print, or publish, or cause to be made, printed, or published any notice, statement, or advertisement, with respect to the sale or rental of a dwelling that indicates any preference, limitation, or discrimination based on race, color, religion, sex, handicap, familial status, or national origin, or an intention to make any such preference, limitation or discrimination.

(d) To represent to any person because of race, color, religion, sex, handicap, familial status, or national origin that any dwelling is not available for inspection, sale, or rental when such swelling is in fact so available.

(e) For profit, to induce or attempt to induce any person to sell or rent any dwelling by representations regarding the entry or prospective entry into the neighborhood of a person or persons of a particular race, color, religion, sex, handicap, familial status, or national origin.

Source: U.S.C. Title 42, Chapter 45, p. 859 et seq., 3601 et seq.

National Environmental Policy Act of 1969

The cornerstone of American environmental policy is the National Environmental Policy Act of 1969, signed by President Richard Nixon on 1 January 1970. In Subchapter I of the act, cited below, the Congress states its position on the nation's environmental policy. Then, in Executive order 11514, President Nixon made his own statement on this topic.

Chapter 55—National Environmental Policy

Subchapter I—Policies and Goals

4331. Congressional declaration of national environmental policy

(a) The Congress, recognizing the profound impact of man's activity on the interrelations of all components of the natural environment, particularly the profound influences of population growth, high-density urbanization, industrial expansion, resource exploitation, and new and expanding technological advances and recognizing further the critical importance of restoring and maintaining environmental quality to the overall welfare and development of man, declares that it is the continuing policy of the Federal Government, in cooperation with State and local governments, and other concerned public and private organizations, to use all practicable means and measures, including financial and technical assistance, in a manner calculated to foster and promote the general welfare, to create and maintain conditions under which man and nature can exist in productive harmony, and fulfill the social, economic, and other requirements of present and future generations of Americans.

(b) In order to carry out the policy set forth in this chapter, it is the continuing responsibility of the Federal Government to use all practicable means, consistent with other essential considerations of national policy, to improve and coordinate Federal plans, functions, programs, and resources to the end that the Nation may—

(1) fulfill the responsibilities of each generation as trustee of the environment for succeeding generations;

(2) assure for all Americans safe, healthful, productive, and esthetically and culturally pleasing surroundings;

(3) attain the widest range of beneficial uses of the environmental without degradation, risk to health or safety, or other undesirable and unintended consequences;

(4) preserve important historic, cultural, and natural aspects of our national heritage, and maintain, wherever possible, an environment which supports diversity and variety of individual choice;

(5) achieve a balance between population and resource use which will permit high standards of living and a wide sharing of life's amenities; and

(6) enhance the quality of renewable resources and approach the maximum attainable recycling of depletable resources.

(c) The Congress recognizes that each person should enjoy a healthful environment and that each person has a responsibility to contribute to the preservation and enhancement of the environment.

4332. Cooperation of agencies; reports; availability of information; recommendations; international and national coordination of efforts

The Congress authorizes and directs that, to the fullest extent possible: (1) the policies, regulations, and public laws of the United States shall be interpreted and administered in accordance with the policies set forth in this chapter and (2) all agencies of the Federal Government shall—

(A) utilize a systematic, interdisciplinary approach which will insure the integrated use of the natural and social sciences and the environmental design arts in planning and in decision-making which may have an impact on man's environment;

(B) identify and develop methods and procedures, in consultation with the Council on Environmental Quality established by subchapter II of this chapter, which will insure that presently unquantified environmental amenities and values may be given appropriate consideration in decision-making along with economic and technical considerations;

(C) include in every recommendation or report on proposals for legislation and other major Federal actions significantly affecting the quality of the human environment, a detailed statement by the responsible official on—

(i) the environmental impact of the proposed action,

(ii) any adverse environmental effects which cannot be avoided should the proposal be implemented,

(iii) alternatives to the proposed action,

(iv) the relationship between local short-term uses of man's environment and the maintenance and enhancement of long-term productivity.

(v) any irreversible and irretrievable commitments of resources which would be involved in the proposed action should it be implemented.

Prior to making any detailed statement, the responsible Federal official shall consult with and obtain the comments of any Federal agency which has jurisdiction by law or special expertise with respect to any environmental impact involved. Copies of such statement and the comments and views of the appropriate Federal, State, and local agencies which are authorized to develop and enforce environmental standards, shall be made available to the President, the Council on Environmental Quality and to the public as provided by section 552 of title 5, and shall accompany the proposal through the existing agency review processes;

(D) Any detailed statement required under subparagraph (C) after January 1, 1970, for any major Federal action funded under a program of grants to States shall not be deemed to be legally insufficient solely by reason of having been prepared by a State agency or official, if:

(i) the State agency or official has statewide jurisdiction and has the responsibility for such action,

(ii) the responsible Federal official furnishes guidance and participates in such preparation,

(iii) the responsible Federal official independently evaluates such statement prior to its approval and adoption, and

(iv) after January 1, 1976, the responsible Federal official provides early notification to, and solicits the view of, any other State or any Federal land management entity of any action or any alternative thereto which may have significant impacts upon State or affected Federal land management entity and, if there is any disagreement on such impacts, prepares a written assessment of such impacts and views for incorporation into such detailed statement.

The procedures in this subparagraph shall not relieve the Federal official of his responsibilities for the scope, objectivity, and content of the entire statement or of any other responsibility under this chapter; and further, this subparagraph does not affect the legal sufficiency of statements prepared by State agencies with less than statewide jurisdiction,

(E) study, develop, and describe appropriate alternatives to recommended courses of action in any proposal which involves unresolved conflicts concerning alternative uses of available resources;

(F) recognize the worldwide and long-range character of environmental problems and, where consistent with the foreign policy of the United States, lend appropriate support to initiatives, resolutions, and programs designed to maximize international cooperation in anticipating and preventing a decline in the quality of mankind's world environment;

(G) make available to States, counties, municipalities, institutions, and individuals, advice and information useful in restoring, maintaining, and enhancing the quality of the environment;

(H) initiate and utilize ecological information in the planning and development of resource-oriented projects; and

(I) assist the Council on Environmental Quality established by subchapter II of this chapter.

Source: U.S.C. Title 42, Chapter 55, p. 963 et seq., 4321 et seq.

Executive Order No. 11514. Protection and Enhancement of Environmental Quality

By virtue of the authority vested in me as President of the United States and in furtherance of the purpose and policy of the National Environmental Policy Act of 1969 (Public Law No. 91-190, approved January 1, 1970), it is ordered as follows:

Sec. 1. Policy

The Federal Government shall provide leadership in protecting and enhancing the quality of the Nation's environment to sustain and enrich human life. Federal agencies shall initiate measures needed to direct their policies, plans and programs so as to meet national environmental goals. The Council on Environmental Quality, through the Chairman, shall advise and assist the President in leading this national effort.

Sec. 2. Responsibilities of Federal Agencies

Consonant with Title I of the National Environmental Policy Act of 1969, hereafter referred to as the "Act," the heads of Federal agencies shall:

(a) Monitor, evaluate, and control on a continuing basis their agencies' activities so as to protect and enhance the quality of the environment. Such activities shall include those directed to controlling pollution and enhancing the environment and those designed to accomplish other program objectives which may affect the quality of the environment. Agencies shall develop programs and measures to protect and enhance environmental quality and shall assess progress in meeting the specific objectives of such activities. Heads of agencies shall consult with appropriate Federal, State and local agencies in carrying out their activities as they affect the quality of the environment.

(b) Develop procedures to ensure the fullest practicable provision of timely public information and understanding of Federal plans and programs with environmental impact in order to obtain the views of interested parties. These procedures shall include, whenever appropriate, provision for public hearings, and shall provide the public with relevant information including information on alternative courses of action. Federal agencies shall also encourage State and local agencies to adopt similar procedures for informing the public concerning their activities affecting the quality of the environment.

(c) Insure that information regarding existing or potential environmental problems and control methods developed as part of research, development, demonstration, test, or evaluation activities is made available to Federal agencies,

States, counties, municipalities, institutions, and other entities, as appropriate.

(d) Review their agencies' statutory authority, administrative regulations, policies, and procedures, including those relating to loans, grants, contracts, leases, licenses, or permits, in order to identify any deficiencies or inconsistencies therein which prohibit or limit full compliance with the purposes and provisions of the Act. A report on this review and the corrective actions taken or planned, including such measures to be proposed to the President, as may be necessary to bring their authority and policies into conformance with the intent, purposes, and procedures of the Act, shall be provided to the Council on Environmental Quality not later than September 1, 1970.

(e) Engage in exchange of data and research results, and cooperate with agencies of other governments to foster the purposes of the Act.

(f) Proceed, in coordination with other agencies, with actions required by section 102 of the Act ["Cooperation of agencies; reports; availability of information; recommendations; international and national coordination of efforts"].

(g) In carrying out their responsibilities under the Act and this Order, comply with the regulations issued by the Council except where such compliance would be inconsistent with statutory requirements.

Sec. 3. Responsibilities of Council on Environmental Quality

The Council on Environmental Quality was established in Subchapter II of the National Environmental Policy Act. In Section 3 of the executive order, President Nixon outlines a number of specific responsibilities of the council and outlines the manner in which the council is to operate.

Sec. 4 Amendments of E.O. 11472

Executive Order No. 11472 concerned the formation of a Cabinet Committee on the Environment, which the Council on Environmental Quality replaces. Section 4 makes changes in terminology of the earlier executive order to bring it into consonance with the new Environmental Policy Act.

Source: Executive Order No. 11514, issued 5 March 1970.

Statutory Civil Rights Requirements of the Environmental Protection Agency

Complaints have been lodged that problems of environmental inequity have at times been caused or aggravated by actions, or lack of actions, taken by the Environmental Protection Agency itself. The point has been made that the EPA, like other federal agencies, is required to operate in accordance with certain general principles, some dealing specifically with civil rights and nondiscrimination considerations. The selections below outline some of the rules and regulations by which the EPA is expected to abide.

7—Nondiscrimination in Programs Receiving Federal Assistance from the Environmental Protection Agency

Subpart A—General
7.10 Purpose of this part
This part implements: Title VI of the Civil Rights Act of 1964, as amended; section 504 of the Rehabilitation Act of 1973, as amended; and section 13 of the Federal Water Pollution Control Act Amendments of 1972, Pub. L. 92-500, (collectively, the Acts).
7.15 Applicability
This part applies to all applicants for, and recipients of, EPA assistance in the operation of programs or activities receiving such assistance beginning February 13, 1984 . . .

Subpart B—Discrimination Prohibited on the Basis of Race, Color, National Origin or Sex
7.30 General prohibition
No person shall be excluded from participation in, be denied the benefits of, or be subject to discrimination under any program or activity receiving EPA assistance on the basis of race, color, national origin, or on the basis of sex in any program or activity receiving EPA assistance under the Federal Water Pollution Control Act, as amended, including the Environmental Finance Act of 1972.
7.35 Specific prohibitions
(a) As to any program or activity receiving EPA assistance, a recipient shall not directly or through contractual, licensing, or other arrangements on the basis of race, color, national origin or, if applicable, sex:

(1) Deny a person any service, aid or other benefit of the program;

(2) Provide a person any service, aid or other benefit that is different, or is provided differently from that provided to others under the program;

(3) restrict a person in any way in the enjoyment of any advantage or privilege enjoyed by others receiving any service, aid, or benefit provided by the program;

(4) Subject a person to segregation in any manner or separate treatment in any way related to receiving services or benefits under the program;

(5) Deny a person or any group of persons the opportunity to participate as members of any planning or advisory body which is an integral part of the program, such as a local sanitation board or sewer authority;

(6) Discriminate in employment on the basis of sex in any program subject to section 13, or on the basis of race, color, or national origin in any program whose purpose is to create employment; or, by means of employment discrimination, deny intended beneficiaries the benefits of the EPA assistance program, or subject the beneficiaries to prohibited discrimination.

(7) In administering a program or activity receiving Federal financial assistance in which the recipient has previously discriminated on the basis of race, color, sex, or national origin, the recipient shall take affirmative action to provide remedies to those who have been injured by the discrimination.

(b) A recipient shall not use criteria or methods of administering its program which have the effect of subjecting individuals to discrimination because of their race, color, national origin, or sex, or have the effect of defeating or substantially impairing accomplishments of the objectives of the program with respect to individuals of a particular race, color, national origin, or sex.

(c) A recipient shall not choose a site or location of a facility that has the purpose or effect of excluding individuals from, denying them the benefits of, or subjecting them to discrimination under any program to which this part applies on the grounds of race, color, or national origin or sex; or with the purpose or effect of defeating or substantially impairing the accomplishment of the objectives of this subpart.

(d) The specific prohibitions of discrimination enumerated above do not limit the general prohibition of 7.30.

Subpart C—Grant Conditions

30.400 General.

All EPA grants are awarded subject to applicable statutory provisions, to requirements imposed pursuant to Executive orders, and to the Grant conditions set forth in this subpart or in Appendix A to this subchapter. Additional special conditions necessary to assure accomplishment of the project or of EPA objectives may be imposed upon any grant or class of grants by agreement with the grantee.

30.401 Statutory conditions.

All EPA grants are awarded subject to the following statutory requirements, in addition to such statutory provision as may be applicable to particular grants or grantees or classes of grants or grantees.

The section then lists statutes to which EPA must adhere in awarding grants, such as the National Environmental Policy Act of 1969, the Clean Air Act, and the Civil Rights Act, as follows:

(c) The Civil Rights Act of 1964, 47 U.S.C. 2000a et seq., as amended and particularly title VI thereof, which provides that no person in the United States shall on the ground of race, color, religion, sex, or national origin be excluded from participation in, be denied the benefits of, or be subject to discrimination under any program or activity receiving Federal financial assistance, as implemented by regulations issued thereunder. . . .

30.402 Executive orders.

All EPA grants are subject to the requirements imposed by the following Executive orders, in addition to such other lawful provisions as may be applicable to particular grants or grantees or classes of grants or grantees.

The section then lists Executive orders to which EPA must adhere in awarding grants, including Executive order 11246, as follows:

(a) Executive Order 11246 (3 CFR, 1964-1965 Comp., p. 339) dated September 24, 1965, as amended, with regard to equal employment opportunities, and all rules, regulations and procedures prescribed pursuant thereto.

Source: 40 CFR Chapter 1, Sections 7 and 30, as cited in *Environmental Justice*. Hearings before the Subcommittee on Civil and Constitutional Rights of the House Committee on the Judiciary, 103rd Congress, 1st Session, 3 and 4 March 1993, pp. 35–40.

Other Environmental Acts

Congress has passed a number of acts dealing with specific aspects of environmental degradation. Some examples include the Clean Air Act of 1970 and later amendments, the Clean Water Act of 1977 and later amendments, the Safe Drinking Water Act of 1974, the Toxic Substances Control Act of 1976, the Solid Waste Disposal Act of 1965, the Resource Recovery Act of 1970, the Resource Conservation and Recovery Act of 1976, the Surface Mining Control and Reclamation Act of 1977, and the Comprehensive Environmental Response, Compensation, and Liability Act of 1980. It is obviously impossible to cite all of these acts. Instead, the following selections review the development of national policy with regard to solid waste disposal.

The first act to deal with such issues was the Solid Waste Disposal Act of 1965. That act focused, as its name suggests, on methods for disposing of solid wastes. Five years later, Congress passed the Resource Recovery Act of 1970 which, for the first time, dealt with recycling useful materials in wastes. In 1970, Congress passed the Resource Conservation and Recovery Act, an omnibus act dealing with all aspects of waste disposal and recovery. Finally, the fourth in this historically important series of acts was the Comprehensive Environmental Response, Compensation and Liability Act, passed in 1980. Perhaps the most significant feature of this act was the establishment of the Superfund, a fund designed to pay for the cleanup of the nation's worst waste disposal sites. Selections from each of these laws are reproduced below.

Solid Waste Disposal Act of 1965

One of the most important environmental laws from the standpoint of environmental justice issues is the Solid Waste Disposal Act of 1965. The reason is that low-income communities and communities of color are frequently confronted with corporate decisions to place hazardous waste dumps within or close to their neighborhoods. The main features of the Solid Waste Disposal Act are outlined in the following section.

Subchapter I—General Provisions

6901. Congressional findings

(a) Solid waste

The Congress finds with respect to solid waste—

(1) that the continuing technological progress and improvement in methods of manufacture, packaging, and marketing of consumer products has resulted in an ever-mounting increase, and in a change in the characteristics, of the mass material discarded by the purchaser of such products;

(2) that the economic and population growth of our Nation, and the improvements in the standard of living enjoyed by our population, have required increased industrial production to meet our needs, and have made necessary the demolition of old buildings, the construction of new buildings, and the provision of highways, and other avenues of transportation, which, together with related industrial, commercial, and agricultural operations, have resulted in a rising tide of scrap, discarded, and waste materials.

(3) that the continuing concentration of our population in expanding metropolitan and other urban areas has presented these communities with serious financial, management, intergovernmental, and technical problems in the disposal of solid wastes resulting from the industrial, commercial, domestic, and other activities carried on in such areas;

(4) that while the collection and disposal of solid wastes should continue to be primarily the function of State, regional, and local agencies, the problems of waste disposal as set forth above have become a matter national in scope and in concern and necessitate Federal action through financial and technical assistance and leadership in the development, demonstration, and application of new and improved methods and processes to reduce the amount of waste and unsalvageable materials and to provide for proper and economical solid waste disposal practices.

(b) Environment and health

The Congress finds with respect to the environment and health, that—

(1) although land is too valuable a national resource to be needlessly polluted by discarded materials, most solid waste is disposed of on land in open dumps and sanitary landfills;

(2) disposal of solid waste and hazardous waste in or on

the land without careful planning and management can present a danger to human health and the environment;

(3) as a result of the Clean Air Act [42 U.S.C. 7401 et seq.], the Water Pollution Control Act [33 U.S.C. 1251 et seq.] and other Federal and State laws respecting public health and the environment, greater amounts of solid waste (in the form of sludge and other pollution treatment residues) have been created. Similarly, inadequate and environmentally unsound practices for the disposal or use of solid waste have created greater amounts of air and water pollution and other problems for the environment and for health;

(4) open dumping is particularly harmful to health, contaminates drinking water from underground and surface supplies, and pollutes the air and the land;

(5) the placement of inadequate controls on hazardous waste management will result in substantial risks to human health and the environment;

(6) if hazardous waste management is improperly performed in the first instance, corrective action is likely to be expensive, complex, and time consuming;

(7) certain classes of land disposal facilities are not capable of assuring long-term containment of certain hazardous wastes, and to avoid substantial risk to human health and the environment, reliance on land disposal should be minimized or eliminated, and land disposal, particularly landfill and surface impoundment, should be the least favored method for managing hazardous wastes; and

(8) alternatives to existing methods of land disposal must be developed since many of the cities in the United States will be running out of suitable solid waste disposal sites within five years unless immediate action is taken.

6902. Objectives and national policy

(a) Objectives

The objectives of this chapter are to promote the protection of health and the environment and to conserve valuable material and energy resources by—

(1) providing technical and financial assistance to State and local governments and interstate agencies for the development of solid waste management plans (including resource recovery and resource conservation systems) which will promote improved solid waste management techniques (including more effective organizational arrangements),

new and improved methods of collection, separation, and recovery of solid waste, and the environmentally safe disposal of nonrecoverable residues;

(2) providing training grants in occupations involving the design, operation, and maintenance of solid waste disposal systems;

(3) prohibiting future open dumping on the land and requiring the conversion of existing open dumps to facilities which do not pose a danger to the environment or to health;

(4) assuring that hazardous waste management practices are conducted in a manner which protects human health and the environment;

(5) requiring that hazardous waste be properly managed in the first instance thereby reducing the need for corrective action at a future date;

(6) minimizing the generation of hazardous waste and the land disposal of hazardous waste by encouraging process substitution, materials recovery, properly conducted recycling and reuse, and treatment;

(7) establishing a viable Federal-State partnership to carry out the purposes of this chapter and insuring that the Administrator will, in carrying out the provisions of subchapter III of this chapter, give a high priority to assisting and cooperating with States in obtaining full authorization of State programs under subchapter III of this chapter.

(8) providing for the promulgation of guidelines for solid waste collection, transport, separation, recovery, and disposal practices and systems;

(9) promoting a national research and development program for improved solid waste management and resource conservation techniques, more effective organizational arrangements, and new and improved methods of collection, separation, and recovery, and recycling of solid wastes and environmentally safe disposal of nonrecoverable residues;

(10) promoting the demonstration, construction, and application of solid waste management, resource recovery, and resource conservation systems which preserve and enhance the quality of air, water, and land resources; and

(11) establishing a cooperative effort among the Federal, State, and local governments and private enterprise in order to recover valuable materials and energy from solid waste.

(b) National policy

The Congress hereby declares it to be the national policy of the United States that, wherever feasible, the generation of hazardous waste is to be reduced or eliminated as expeditiously as possible. Waste that is nevertheless generated should be treated, stored, or disposed of so as to minimize the present and future threat to human health and the environment.

Comprehensive Environmental Response, Compensation, and Liability Act of 1980

Federal policy and practices with regard to solid waste disposal were updated in the Comprehensive Environmental Response, Compensation, and Liability Act of 1980. This act created, among other things, a Superfund designed to pay for the cleanup of hazardous waste sites. Some passages of potential interest with respect to environmental inequities are cited below.

9602. Designation of additional hazardous substances and establishment of reportable released quantities; regulations

(a) The Administrator shall promulgate and revise as may be appropriate, regulations designating as hazardous substances, in addition to those referred to in section 9601(14) of this title, such elements, compounds, mixtures, solutions, and substances which when released into the environment may present substantial danger to the public health or welfare of the environment, and shall promulgate regulations establishing that quantity of any hazardous substance the released of which shall be reported pursuant to section 9603 of this title. The Administrator may determine that one single quantity shall be the reportable quantity for any hazardous substance, regardless of the medium into which the hazardous substance is released. For all hazardous substances for which proposed regulations establishing reportable quantities were published in the Federal Register under this subsection on or before March 1, 1986, the Administrator shall promulgate under this subsection final regulations establishing reportable quantities not later than December 31, 1986. For all hazardous substances for which proposed regulations establishing reportable quantities were not published in the Federal Register under this subsection on or before March 1, 1986, the Administrator shall publish under this subsection proposed regulations establishing reportable quantities not later than December 31, 1986, and promulgate final regulations under this subsection establishing reportable quantities not later than April 30, 1986." [sic]

(b) Unless and until superseded by regulations establishing a reportable quantity under subsection (a) of this section for any hazardous substance as defined in section 9601(14) of this title, (1) a quantity of one pound or (2) for those hazardous substances for which reportable quantities have been established pursuant to section 1321(b)(4) of title 33, such reportable quantity, shall be deemed that quantity, the release of which requires notification pursuant to section 9603(a) or (b) of this title.

9603. Notification requirements respecting released substance . . .

(c) Notice to Administrator of EPA of existence of storage, etc., facility by owner or operator; exception; time, manner, and form of notice; penalties for failure to notify; use of notice of information pursuant to notice in criminal case.

Within one hundred and eighty days after December 11, 1980, any person who owns or operates or who at the time of disposal owned or operated, or who accepted hazardous substances for transport and selected, a facility at which hazardous substances (as defined in section 9601(14)(C) of this title) are or have been stored, treated, or disposed of shall, unless such facility has a permit issued under, or has been accorded interim status under, subtitle C of the Solid Waste Disposal Act [42 U.S.C. 6921 et seq.], notify the Administrator of the Environmental Protection Agency of the existence of such facility, specifying the amount and type of any hazardous substance to be found there, and any known, suspect, or likely releases of such substances from such facility. The Administrator may prescribe in greater detail the manner and form of the notice and the information included. The Administrator shall notify the affected State agency, or any department designated by the Governor to receive such notice, of the existence of such facility. Any person who knowingly fails to notify the Administrator of the existence of any such facility shall, upon conviction, be fined not more than $10,000, or imprisoned for not more than one year, or both. In addition, any such person who knowingly fails to provide the notice required by this subsection shall not be entitled to any limitation or liability or to any defenses to liability set out in section 9607 of this title: *provided, however,* that notification under this subsection is not required for any facility which would be reportable hereunder solely as a result of any stoppage in transit which is temporary, incidental to the transportation movement, or at the ordinary operating convenience of a common or contract carrier, and such stoppage shall be considered as a continuity of movement and not as the storage of a hazardous substance. Notification received pursuant to this subsection or information obtained by the exploitation of such notification shall not be used against such person in any criminal case, except a prosecution for perjury or for giving a false statement.

(d) Recordkeeping requirements; promulgation of rules and regulations by Administrator of EPA; penalties for violations; waiver of retention requirements.

(1) The Administrator of the Environmental Protection Agency is authorized to promulgate rules and regulations specifying, with respect to—

(A) the location, title, or condition of a facility, and

(B) the identity, characteristics, quantity, origin, or condition (including containerization and previous treatment) of any hazardous substances contained or deposited in a facility; the records which shall be retained by any person required to provide the notification of a facility set out in subsection (c) of this section. Such specification shall be in accordance with the provisions of this subsection. . . .

9604. Response authorities

(a) Removal and other remedial action by President; applicability of national contingency plan; response by potentially responsible parties; public health threats; limitations on response; exception

(1) Whenever (A) any hazardous substance is released or there is a substantial threat of such a release into the environment, or (B) there is a release or substantial threat of release into the environment of any pollutant or contaminant which may present an imminent and substantial danger to the public health or welfare, the President is authorized to act, consistent with the national contingency plan, to remove or arrange for the removal of, and provide for remedial action relating to such hazardous substance, pollutant, or

contaminant at any time (including its removal from any contaminated natural resource), or take any other response measure consistent with the national contingency plan which the President deems necessary to protect the public health or welfare or the environment. When the President determines that such action will be done properly and promptly by the owner or operator of the facility of vessel or by any other responsible party, the President may allow such person to carry out the action, conduct the remedial investigation, or conduct the reliability study in accordance with section 9622 of this title. No remedial investigation or feasibility study (RI/FS) shall be authorized except on a determination by the President that the party is qualified to conduct the RI/FS and only if the President contracts with or arranges for a qualified person to assist the President in overseeing and reviewing the conduct of such RI/FS and if the responsible party agrees to reimburse the [Super]Fund for any cost incurred by the President under, or in connection with, the oversight contract or arrangement. In no event shall a potentially responsible party be subject to a lesser standard of liability, receive preferential treatment, or in any other way, whether direct or indirect, benefit from any such arrangements as a response action contractor, or as a person hired or retained by such a response action contractor, with respect to the release or facility in question. The President shall give primary attention to those releases which the President deems may present a public health threat . . .

(b) Investigations, monitoring, coordination, etc., by President

(1) Information; studies and investigations

Whenever the President is authorized to act pursuant to subsection (a) of this section, or whenever the President has reason to believe that a release has occurred or is about to occur, or that illness, disease, or complaints thereof may be attributable to exposure to a hazardous substance, pollutant, or contaminant and that a release may have occurred or by occurring, he may undertake such investigations, monitoring, surveys, testing, and other information gathering as he may deem necessary or appropriate to identify the existence and extent of the release or threat thereof, the source and nature of the hazardous substances, pollutants or contaminants involved, and the extent of danger to the public health or welfare or to the environment. In addition, the President may undertake such planning, legal, fiscal, economic, engineering, architectural, and other studies or investigations as he may deem necessary or appropriate to plan and direct response actions, to recover the costs thereof and to enforce the provisions of this chapter.

Source: U.S.C. Title 42, Chapter 103, p. 844 et seq., 9601 et seq.

Resource Conservation and Recovery Act

One of the difficulties in dealing with problems of environmental inequity is that many of the laws stated above apply only to federal projects, while many of the issues of environmental inequity that arise fall under state jurisdiction. Since only two states (Arkansas and Florida) have passed environmental justice laws, plaintiffs may be left with little or no legal remedy for problems of disproportionate exposure to hazardous waste sites. The selections from the Resource Conservation and Recovery Act reproduced below suggest, however, another way of dealing with such problems. Since the federal government funds many of the waste management activities actually operating under state control, it may legitimately impose restrictions, such as those provided by the Civil Rights Act of 1964, on states and corporations that receive federal funding.

6947. Approval of State plan; Federal assistance

(a) Plan approval

The Administrator shall, within six months after a State plan has been submitted for approval, approve or disapprove the plan. The Administrator shall approve a plan if he determines that—[provisions for approving or disapproving a plan are described.]

The Administrator shall review approved plans from time to time and if he determines that revision or corrections are necessary to bring such plan into compliance with the minimum requirements promulgated under section 6943 of this title (including new or revised requirements), he shall, after notice and opportunity for public hearing, withdraw his approval of such plan. Such withdrawl [sic] of approval shall cease to be effective upon the Administrator's determination that such complies with such minimum requirements.

(b) Eligibility of States for Federal financial assistance

(1) The Administrator shall approve a State application

for financial assistance under this subchapter, and make grants to such State, if such State and local and regional authorities within such State have complied with the requirements of section 6946 of this title within the period required under such section and if such State has a State plan which has been approved by the Administrator under this subchapter.

(2) The Administrator shall approve a State application for financial assistance under this subchapter, and make grants to such State, for fiscal years 1978 and 1979 if the Administrator determines that the State plan continues to be eligible for approval under subsection (a) of this section and is being implemented by the State.

(3) Upon withdrawl [sic] of approval of a State plan under subsection (a) of this section, the Administrator shall withhold Federal financial and technical assistance under this subchapter (other than such technical assistance as may be necessary to assist in obtaining the reinstatement of approval) until such time as such approval is reinstated . . .

6948. Federal assistance
(a) Authorization of Federal financial assistance

This section authorizes specific amounts of monies for financial assistance to states in 1978 through 1988.

(2)(A) The Administrator is authorized to provide financial assistance to States, counties, municipalities, and intermunicipal agencies and State and local public solid waste management authorities for implementation of programs to provide solid waste management, resource recovery, and resource conservation services and hazardous waste management. Such assistance shall include assistance for facility planning and feasibility studies; expert consultation; surveys and analyses of market needs; marketing of recovered resources; technology assessments; legal expenses; construction feasibility studies; source separation projects; and fiscal or economic investigations or studies; but such assistance shall not include any other element of construction, or any acquisition of land or interest in land, or any subsidy for the price of recovered resources. Agencies assisted under this subsection shall consider existing solid waste management and hazardous waste management services and facilities as well as facilities proposed for construction.

(B) An applicant for financial assistance under this paragraph must agree to comply with respect to the project or program assisted with the applicable requirements of section 6945 of this title and subchapter III of this chapter and apply applicable solid waste management practices, methods, and levels of control consistent with any guidelines published pursuant to section 6907 of this title. Assistance under this paragraph shall be available only for programs certified by the State to be consistent with any applicable State or areawide solid waste management plan or program. Applicants for technical and financial assistance under this section shall not preclude or foreclose consideration of programs for the recovery of recyclable material through source separation or other resource recovery techniques.

This section then concludes with other subsections dealing with related matters, such as:

(b) State allotment . . .
(c) Distribution of Federal financial assistance with the State . . .
(d) Technical assistance . . .
(e) Special communities . . .
(f) Assistance to States for discretionary program for recycled oil . . .
(g) Assistance to municipalities for energy and materials conservation and recovery planning activities . . .

6949. Rural communities assistance
(a) In general
The Administrator shall make grants to States to provide assistance to municipalities with a population of five thousand or less, or counties with a population of ten thousand or less or less than twenty persons per square mile and not within a metropolitan area, for solid waste management facilities (including equipment) necessary to meet the requirements of section 6945 of this title or restrictions on open burning or other requirements arising under the Clean Air Act [42 U.S.C. [7401 et seq.] or the Federal Water Pollution Control Act [33 U.S.C. 1251 et seq.] . . .

A variety of conditions pertaining to such grants is then outlined.

6949a. Adequacy of certain guidelines and criteria
(a) Study
The Administrator shall conduct a study of the extent to

which the guidelines and criteria under this chapter (other than guidelines and criteria for facilities to which subchapter III of this chapter applies) which are applicable to solid waste management and disposal facilities, including, but not limited to landfills and surface impoundments, are adequate to protect human health and the environment from ground water contamination. Such study shall include a detailed assessment of the degree to which the criteria under section 6907(a) of this title and the criteria under section 6944 of this title regarding monitoring, prevention of contamination, and remedial action are adequate to protect ground water and shall also include recommendation with respect to any additional enforcement authorities which the Administrator, in consultation with the Attorney General, deems necessary for such purposes.

Source: U.S.C. Title 42, Chapter 82, page 435 et seq., 6947–6949.

Toxic Substances Control Act of 1976

An act dealing somewhat more broadly with the manufacture, processing, distribution, consumption, and disposal of hazardous materials was the Toxic Substances Control Act of 1976. Along with the waste disposal acts cited above, the Toxic Substances Control Act of 1976 may be relevant in some issues of environmental inequity. Some pertinent sections from that act are reproduced below.

Subchapter I—Control of Toxic Substances

2601.Findings, policy, and intent

(a) Findings

The Congress finds that—

(1) human beings and the environment are being exposed each year to a large number of chemical substances and mixtures;

(2) among the many chemical substances and mixtures which are constantly being developed and produced, there are some whose manufacture, processing, distribution in commerce, use, or disposal may present an unreasonable risk of injury to health or the environment; and

(3) the effective regulation of interstate commerce in such chemical substances and mixtures also necessitates the regulation of intrastate commerce in such chemical substances and mixtures.

(b) Policy

It is the policy of the United States that—

(1) adequate data should be developed with respect to the effect of chemical substances and mixtures on health and the environment and that the development of such data should be the responsibility of those who manufacture and those who process such chemical substances and mixtures;

(2) adequate authority should exist to regulate chemical substances and mixtures which present an unreasonable risk of injury to health or the environment, and to take action with respect to chemical substances and mixtures which are imminent hazards; and

(3) authority over chemical substances and mixtures should be exercised in such a manner as not to impede unduly or create unnecessary economic barriers to technological innovation while fulfilling the primary purpose of this chapter to assure that such innovation and commerce in such chemical substances and mixtures do not present an unreasonable risk of injury to health or the environment.

(c) Intent of Congress

It is the intent of Congress that the Administrator shall carry out this chapter in a reasonable and prudent manner, and that the Administrator shall consider the environmental, economic, and social impact of any action the Administrator takes or proposes to take under this chapter. . . .

2619. Citizens' civil actions

(a) In general

Except as provided in subsection (b) of this section, any person may commence a civil action—

(1) against any person (including (A) the United States, and (B) any other governmental instrumentality or agency to the extent permitted by the eleventh amendment to the Constitution) who is alleged to be in violation of this chapter or any rule promulgated under section 2603, 2604 or 2605 of this title, or subchapter II of this chapter, or order issued under section 2604 of this title or subchapter II of this chapter to restrain such violation, of

(2) against the Administrator to compel the Administrator to perform any act or duty under this chapter which is not discretionary. Any civil action under paragraph (1) shall be brought in the United States district court for the district in which the alleged violation occurred or in which the

defendant resides or in which the defendant's principal place of business is located. Any action brought under paragraph (2) shall be brought in the United States District Court for the District of Columbia, or the United States district court for the judicial district in which the plaintiff is domiciled. The district courts of the United States shall have jurisdiction over suits brought under this section, without regard to the amount in controversy or the citizenship of the parties. In any civil action under this subsection process may be served on a defendant in any judicial district in which the defendant resides or may be found and subpoenas for witnesses may be served in any judicial district. . . .

2620. Citizens' petitions

(a) In general

Any person may petition the Administrator to initiate a proceeding for the issuance, amendment, or repeal of a rule under section 2603, 2605, or 2607 of this title or an order under section 2604(e) or 2605(b)(2) of this title. . . .

2622. Employee protection

(a) In general

No employer may discharge any employee or otherwise discriminate against any employee with respect to the employee's compensation, terms, conditions, or privileges or employment because the employee (or any person acting pursuant to a request of the employee) has—

(1) commenced, caused to be commenced, or is about to commence or cause to be commenced a proceeding under this chapter;

(2) testified or is about to testify in any such proceeding; or

(3) assisted or participate or is about to assist or participate in any manner in such a proceeding or in any other action to carry out the purposes of this chapter.

Source: U.S.C. Title 15, Chapter 53, p. 1231 et seq., 2601 et seq.

The Environmental Equal Rights Act of 1993

The Environmental Equal Rights Act was first introduced in 1992 under the sponsorship of John Lewis (D-Georgia) and Al Gore (D-Tennessee). It was not acted upon and was reintroduced a year later by Lewis and Max Baucus (D-Montana).

H.R. 1924

A Bill

To amend the Solid Waste Disposal Act to allow petitions to be submitted to prevent certain waste facilities from being constructed in environmentally disadvantaged communities.

Be it enacted by the Senate and House of Representatives of the United States of America in Congress assembled,

Sec. 1. Short Title

This Act may be cited as the "Environmental Equal Rights Act of 1993".

Sec. 2. Findings

The Congress finds the following:

(1) A 1987 study by the United Church of Christ found that the proportion of minorities in communities with large commercial landfills or a high number of commercial waste facilities was 3 times greater than in communities without such facilities.

(2) The same United Church of Christ study found that approximately 60 percent of African- and Hispanic-Americans live in a community that has an uncontrolled hazardous waste site.

(3) An Environmental Protection Agency report released in 1992 found that racial minority and low-income populations experience higher than average exposure to selected air pollutants and hazardous waste facilities.

(4) A 1983 analysis by the General Accounting Office found that, in the southwestern United States, 3 of the 4 commercial hazardous waste landfills were located in communities with more blacks than whites, and the percentage of residents near the sites with incomes below the poverty line ranged from 26 percent to 42 percent.

(5) A University of Michigan study released in 1990 found that minorities were 4 times more likely than whites to live within 1 mile of a commercial hazardous waste facility in the 3-county Detroit metropolitan area.

(6) A National Law Journal study found that penalties

imposed for pollution law violations in areas predominantly populated by minorities were dramatically lower than those imposed for violations in large white areas.

Sec. 3. Petition Relating to Environmentally Disadvantaged Communities

(a) AMENDMENT TO SUBTITLE G.—Subtitle G of the Solid Waste Disposal Act (42 U.S.C. 6971 et seq.) is amended by adding at the end the following new section:
"SEC. 7014. PETITION RELATING TO ENVIRONMEN-TALLY DISADVANTAGED COMMUNITIES.

"(a) RIGHT TO PETITION.—(1) Any citizen residing in a State in which a new facility for the management of solid waste (including a new facility for the management of hazardous waste) is proposed to be constructed in an environmentally disadvantaged community may submit a petition to the appropriate entity (described in paragraph (2)) to prevent the proposed facility from being issued a permit to be constructed or to operate in that community.

"(2) A petition under paragraph (1) shall be submitted in accordance with the following subparagraphs:

"(A) In the case of a facility for the management of hazardous waste, the petition shall be submitted to the Administrator or, in the case of a State with an authorized program under section 3006, to the State.

"(B) In the case of a facility for the management of municipal solid waste, the petition shall be submitted to the Administrator or, in appropriate cases, as determined under regulations implementing this section, to the State.

"(b) AGENCY HEARING—(1) Within a reasonable period of time after receipt of a petition under subsection (a), the Administrator or the State shall hold a public hearing on the petition. An administrative law judge of the Environmental Protection Agency or an equivalent employee of the State, in the case of a petition submitted to the State, shall preside at the hearing.

"(2) Subject to paragraph (3), the administrative law judge or State employee shall approve the petition if, at the hearing, the petitioner established that—

"(A) the proposed facility will be located in an environmentally disadvantaged community; and

"(B) the proposed facility may adversely affect—

"(i) the human health of such community or a portion of such community; or

"(ii) the air, soil, water, or other elements of the community or a portion of such community.

"(3) After the petitioner has satisfied the requirement of paragraph (2), the administrative law judge or State employee shall deny the petition only if, at the hearing, the proponent of the proposed facility establishes that—

"(A) there is no alternative location within the State for the proposed facility that poses fewer risks to human health and the environment than the proposed facility (according to standards for comparing the degree of risk to human health and the environment promulgated in regulations by the Administrator for purposes of this section); and

"(B) the proposed facility—

"(i) will not release contaminants; or

"(ii) will not engage in any activity that is likely to increase the cumulative impact of contaminants on any residents of the environmentally disadvantaged community.

"(c) ADMINISTRATIVE PROVISIONS—(1) The submission of a petition under subsection (a) stays the issuance of a permit for the facility concerned until a decision on the petition has been rendered under subsection (b).

"(2) If more than one petition relating to the same facility is submitted, the petitions may be consolidated by the appropriate official to promote the efficient resolution and disposition of the petitions.

"(d) DEFINITIONS—For purposes of this section:

"(1) The term 'environmentally disadvantaged community' means an area within 2 miles of the borders of a site on which a facility for the management of solid waste (including a facility for the management of hazardous waste) is proposed to be constructed and in which both of the following conditions are met, determined using the most recent data from the Bureau of the Census:

"(A)(i) The percentage of the population consisting of all individuals who are of African, Hispanic, Asian, Native American Indian, Pacific Island, or Native Alaskan ancestry is greater than either—

"(I) the percentage of the population in the State of such individuals, or

"(II) the percentage of the population in the United States of all such individuals; or "(ii)(I) twenty percent or more of the population consists of individuals who are living at or below the poverty line, or

"(II) the area has a per capita income of 80 percent or less of the national average, for the most recent 12-month period for which statistics are available.

"(B) The area contains one or more of the following:

"(i) A facility for the management of hazardous waste that is in operation.

"(ii) A facility for the management of hazardous waste that is no longer in operation but that formerly accepted hazardous waste.

"(iii) A site at which a release or threatened release of hazardous substances (within the meaning of the Comprehensive Environmental Response, Compensation, and Liability Act of 1980) has occurred.

"(iv) A facility for the management of municipal solid waste.

"(v) A facility whose owner or operator is required to submit a toxic chemical release form under section 313 of the Emergency Planning and Community Right-to-Know Act of 1986 (42 U.S.C. 11023), if the releases reported on such form are likely to adversely affect the human health of the community or portion of the community, as determined by the entity that would be appropriate under subsection (a)(2) if a petition were filed with respect to the facility.

"(2)" The term 'management', when used in connection with solid waste (including hazardous waste), means treatment, storage, disposal, combustion, recycling, or other handling of solid waste, but does not include any activities that take place in a materials recovery facility or any other facility that prepares, transfers, or utilizes nonhazardous recyclable materials for purposes other than energy recovery.

"(3) the terms 'release' and 'containment' have the meanings prescribed by the Administrator for purposes of this section."

(b) TABLE OF CONTENTS AMENDMENT.—The table of contents for subtitle G of such Act is further amended by adding at the end the following new item:

"Sec. 7014. Petition relating to environmental disadvantaged communities."

Executive Orders

Executive Order 12898 (President Clinton)—Federal Actions To Address Environmental Justice in Minority Populations and Low-Income Populations

February 11, 1994

By the authority vested in me as President by the Constitution and the laws of the United States of America, it is hereby ordered as follows:

Section 1-1. Implementation

1-101. Agency Responsibilities. To the greatest extent practicable and permitted by law, and consistent with the principles set forth in the report on the National Performance Review, each Federal agency shall make achieving environmental justice part of its mission by identifying and addressing, as appropriate, disproportionately high and adverse human health or environmental effects of its programs, policies, and activities on minority populations and low-income populations in the United States and its territories and possessions, the District of Columbia, the Commonwealth of Puerto Rico, and the Commonwealth of the Mariana Islands.

1-102. Creation of an Interagency Working Group on Environmental Justice. (a) Within 3 months of the date of this order, the Administrator of the Environmental Protection Agency ("Administrator") or the Administrator's designee shall convene an interagency Federal Working Group on Environmental Justice ("Working Group"). The Working Group shall comprise the heads of the following executive agencies and offices, or their designees: (a) Department of Defense; (b) Department of Health and Human Services; (c) Department of Housing and Urban Development; (d) Department of Labor; (e) Department of Agriculture; (f) Department of Transportation; (g) Department of Justice; (h) Department of the Interior; (i) Department of Commerce; (j) Department of Energy; (k) Environmental Protection Agency; (l) Office of Management and Budget; (m) Office of Science and Technology Policy; (n) Office of the Deputy Assistant to the President for Environmental Policy; (o) Office of the Assistant to the President for Domestic Policy; (p) National Economic Council; (q) Council of Economic Advisers; and (r) such other Government officials as the President may designate. The Working Group shall report to the President through the Deputy Assistant to the President for Environmental Policy and the Assistant to the President for Domestic Policy.

(b) The Working Group shall: (1) provide guidance to Federal agencies on criteria for identifying disproportionately high and adverse human health or environmental effects on minority populations and low-income populations;

(2) coordinate with, provide guidance to, and serve as a clearinghouse for, each Federal agency as it develops an environmental justice strategy as required by section 1-103 of this order, in order to ensure that the administration, interpretation and enforcement of programs, activities and policies are undertaken in a consistent manner;

(3) assist in coordinating research by, and stimulating cooperation among, the Environmental Protection Agency, the Department of Health and Human Services, the Department of Housing and Urban Development, and other agencies conducting research or other activities in accordance with section 3-3 of this order;

(4) assist in coordinating data collection, required by this order;

(5) examine existing data and studies on environmental justice;

(6) hold public meetings as required in section 5-502(d) of this order; and

(7) develop interagency model projects on environmental justice that evidence cooperation among Federal agencies.

1-103. Development of Agency Strategies. (a) Except as provided in section 6-605 of this order, each Federal agency shall develop an agency-wide environmental justice strategy, as set forth in subsections (b)-(e) of this section that identifies and addresses disproportionately high and adverse human health or environmental effects of its programs, policies, and activities on minority populations and low-income populations. The environmental justice strategy shall list programs, policies, planning and public participation processes, enforcement, and/or rulemakings related to human health or the environment that should be revised to, at a minimum: (1) promote enforcement of all health and environmental statutes in areas with minority populations and low-income populations; (2) ensure greater public participation; (3) improve research and data collection relating to the health of and environment of minority populations and low-income populations; (4) identify differential patterns of consumption of natural resources among minority populations and low-income populations. In addition, the environmental justice strategy shall include, where appropriate, a timetable for undertaking identified revisions and consideration of economic and social implications of the revisions.

(b) Within 4 months of the date of this order, each Federal agency shall identify an internal administrative process for developing its environmental justice strategy, and shall inform the Working Group of that process.

(c) Within 6 months of the date of this order, each Federal agency shall provide the Working Group with an outline of its proposed environmental justice strategy.

(d) Within 10 months of the date of this order, each Federal agency shall provide the Working Group with its proposed environmental justice strategy.

(e) Within 12 months of the date of this order, each Federal agency shall finalize its environmental justice strategy and provide a copy and written description of its strategy to the Working Group. During the 12 month period from the date of this order, each Federal agency, as part of its environmental justice strategy, shall identify several specific projects that can be promptly undertaken to address particular concerns identified during the development of the proposed environmental justice strategy, and a schedule for implementing those projects.

(f) Within 24 months of the date of this order, each Federal agency shall report to the Working Group on its progress in implementing its agency-wide environmental justice strategy.

(g) Federal agencies shall provide additional periodic reports to the Working Group as requested by the Working Group.

1-104. Reports to the President. Within 14 months of the date of this order, the Working Group shall submit to the President, through the Office of the Deputy Assistant to the President for Environmental Policy and the Office of the Assistant to the President for Domestic Policy, a report that describes the implementation of this order, and includes the final environmental justice strategies described in section 1-103(e) of this order.

Sec. 2-2. Federal Agency Responsibilities for Federal Programs. Each Federal agency shall conduct its programs, policies, and activities that substantially affect human health or the environment, in a manner that ensures that such programs, policies, and activities do not have the effect of excluding persons (including populations) from participation in, denying persons (including populations) the benefits of, or subjecting persons (including populations) to discrimination under, such programs, policies, and activities, because of their race, color, or national origin.

Sec. 3-3. Research, Data Collection, and Analysis.

3-301. Human Health and Environmental Research and Analysis. (a) Environmental human health research, whenever practicable and appropriate, shall include diverse segments of the population in epidemiological and clinical studies, including segments at high risk from environmental hazards, such as minority populations, low-income populations and workers who may be exposed to substantial environmental hazards.

(b) Environmental human health analyses, whenever practicable and appropriate, shall identify multiple and cumulative exposures.

(c) Federal agencies shall provide minority populations and low-income populations the opportunity to comment on the development and design of research strategies undertaken pursuant to this order.

3-302. Human Health and Environmental Data Collection and Analysis. To the extent permitted by existing law, including the Privacy Act, as amended (5 U.S.C. section 552a): (a) each Federal agency, whenever practical and appropriate, shall collect, maintain, and analyze information assessing and comparing environmental and human health risks borne by populations identified by race, national origin, or income. To the extent practical and appropriate, Federal agencies shall use this information to determine whether their programs, policies, and activities have disproportionately high and adverse human health or environmental effects on minority populations and low-income populations;

(b) In connection with the development and implementation of agency strategies in section 1-103 of this order, each Federal agency, whenever practicable and appropriate, shall collect, maintain and analyze information on the race, national origin, income level, and other readily accessible and appropriate information for areas surrounding facilities or sites expected to have a substantial environmental human health, or economic effect on the surrounding populations, when such facilities or sites become the subject of a substantial Federal environmental administrative or judicial action. Such information shall be made available to the public, unless prohibited by law; and

(c) Each Federal agency, whenever practicable and appropriate, shall collect, maintain, and analyze information on the race, national origin, income level, and other readily accessible and appropriate information for areas surrounding Federal facilities that are: (1) subject to the reporting requirements under the Emergency Planning and Community Right-to-Know Act, 42 U.S.C. section 11001-11050 as mandated in Executive Order No. 12856; and (2) expected to have a substantial environmental human health, or economic effect on surrounding populations. Such information shall be made available to the public, unless prohibited by law.

(d) In carrying out the responsibilities in this section, each Federal agency, whenever practicable and appropriate, shall share information and eliminate unnecessary duplication of efforts through the use of existing data systems and cooperative agreements among Federal agencies with States, local, and tribal governments.

Sec. 4-4. Subsistence Consumption of Fish and Wildlife.

4-401. Consumption Patterns. In order to assist in identifying the need of ensuring protection of populations with differential patterns of subsistence consumption of fish and wildlife, Federal agencies, whenever practicable and appropriate, shall collect, maintain, and analyze information on the consumption patterns of populations who principally rely on fish and/or wildlife for subsistence. Federal agencies shall communicate to the public the risks of those consumption patterns.

4-402 Guidance. Federal agencies, whenever practicable and appropriate, shall work in a coordinated manner to publish guidance reflecting the latest scientific information available concerning methods for evaluating the human health risks associated with the consumption of pollutant-bearing fish or wildlife. Agencies shall consider such guidance in developing their policies and rules.

Sec. 5-5. Public Participation and Access for Information. (a) The public may submit recommendations to Federal agencies relating to the incorporation of environmental justice principles into Federal agency programs or policies. Each Federal agency shall convey such recommendations to the Working Group.

(b) Each Federal agency may, whenever practicable and appropriate, translate crucial public documents, notices, and hearings relating to human health or the environment for limited English speaking populations.

(c) Each Federal agency shall work to ensure that public documents, notices, and hearings relating to human health or the environment are concise, understandable, and readily accessible to the public.

(d) The Working Group shall hold public meetings, as appropriate, for the purpose of fact-finding, receiving public comments, and conducting inquires concerning environmental justice. The Working Group shall prepare for public review a summary of the comments and recommendations discussed at the public meetings.

Sec. 6-6. General Provisions.

6-601. Responsibility for Agency Implementation. The head of each Federal agency shall be responsible for ensuring compliance with this order. Each Federal agency shall conduct internal reviews and take such other steps as many be necessary to monitor compliance with this order.

6-602. Executive Order No. 12250. This Executive order is intended to supplement but not supersede Executive Order No. 12250, which requires consistent and effective implementation of various laws prohibiting discriminatory practices in programs receiving Federal financial assistance. Nothing herein shall limit the effect or mandate of Executive Order No. 12250.

6-603. Executive Order No. 12875. This Executive order is not intended to limit the effect or mandate of Executive Order No. 1875.

6-604. Scope. For purposes of this order, Federal agency means any agency on the Working Group, and such other agencies as may be designated by the President, that conducts any federal program or activity that substantially affects human health or the environment. Independent agencies are requested to comply with the provisions of this order.

6-605. Petitions for Exemptions. The head of a Federal agency may petition the President for an exemption from the requirements of this order on the grounds that all or some of the petitioning agency's programs or activities should not be subject to the requirements of this order.

6-606. Native American Programs. Each Federal agency responsibility set forth under this order shall apply equally to Native American programs. In addition, the Department of the Interior, in coordination with the Working Group, and after consultation with tribal leaders, shall coordinate steps to be taken pursuant to this order that address Federally-recognized Indian Tribes.

6-607. Costs. Unless otherwise provided by law, Federal agencies shall assume the financial costs of complying with this order.

6-608. General. Federal agencies shall implement this order consistent with, and to the extent permitted by, existing law.

6-609. Judicial Review. This order is intended only to improve the internal management of the executive branch and is not intended to, nor does it create any right, benefit, or trust responsibility, substantive or procedural, enforceable at law or equity by a party against the United States, its agencies, its officers, or any person. This order shall not be construed to create any right to judicial review involving the compliance or non-compliance of the United States, its agencies, its officers, or any other person with this order.

William J. Clinton

The White House
February 11, 1994.

[Filed with the Office of the Federal Register, 3:07 p.m., February 14, 1994]

[Note: This Executive order was published in the Federal Register on February 16.]

Memorandum on Environmental Justice

February 11, 1994

*Memorandum for the Heads
of All Departments and Agencies*

Subject: Executive Order on Federal Actions To Address Environmental Justice in Minority Populations and Low-Income Populations

Today I have issued an Executive order on Federal Actions to Address Environmental Justice in Minority Populations and Low-Income Populations. That order is designed to focus Federal attention on the environmental and human health conditions in minority communities and low-income communities with the goal of achieving environmental justice. That order is also intended to promote nondiscrimination in Federal programs substantially affecting human health and the environment, and to provide minority communities and low-income communities access to public information on, and an opportunity for public participation in, matters relating to human health or the environment.

The purpose of this separate memorandum is to underscore certain provision [sic] of existing law that can help ensure that all communities and persons across this Nation live in a safe and healthful environment. Environmental and civil rights statutes provide many opportunities to address environmental hazards in minority communities and low-income communities. Application of these existing statutory provisions is an important part of this Administration's efforts to prevent those minority communities and low-income communities from being subject to disproportionately high and adverse environmental effects.

I am therefore today directing that all department and agency heads take appropriate and necessary steps to ensure that the following specific directives are implemented immediately:

In accordance with Title VI of the Civil Rights Act of 1964, each Federal agency shall ensure that all programs or activities receiving Federal financial assistance that affect human health or the environment do not directly, or through contractual or other arrangements, use criteria, methods, or practices that discriminate on the basis of race, color, or national origin.

Each Federal agency shall analyze the environmental effects, including human health, economic and social effects, of Federal actions, including effects on minority communities and low-income communities, when such analysis is required by the national Environmental Policy Act of 1969 (NEPA), 42 U.S.C. section 4321 *et seq.* Mitigation measures outlined or analyzed in an environmental assessment, environmental impact statement, or record of decision, whenever feasible, should address significant and adverse environmental effects of proposed Federal actions on minority communities and low-income communities.

Each Federal agency shall provide opportunities for community input in the NEPA process, including identifying potential effects and mitigation measures in consultation with affected communities and improving the accessibility of meetings, crucial documents, and notices.

The Environmental Protection Agency, when reviewing environmental effects of proposed action of other Federal agencies under section 309 of the Clean Air Act, 42 U.S.C. section 7609, shall ensure that the involved agency has fully analyzed environmental effects on minority communities and low-income communities, including human health, social, and economic effects.

Each Federal agency shall ensure that the public, including minority communities and low-income communities, has adequate access to public information relating to human health or environmental planning, regulations, and enforcement when required under the Freedom of Information Act, 5 U.S.C. section 552, the Sunshine Act, 5 U.S.C. section 552b, and the emergency Planning and Community Right-to-Know Act, 42 U.S.C. section 11044.

This memorandum is intended only to improve the internal management of the Executive Branch and is not intended to nor does it create, any right, benefit, or trust responsibility, substantive or procedural, enforceable at law or equity by a party against the United States, its agencies, its officers, or any person.

William J. Clinton

Arkansas State Law on Environmental Equity in Siting High-Impact Solid Waste Management Facilities

As of mid 1996, only two states, Arkansas and Florida, had passed laws dealing with problems of environmental inequities. The Arkansas bill, House Bill 1986, was passed by the General Assembly and Senate and signed by Governor Jim Guy Tucker on 20 April 1993. Italicized portions of the act represent changes that have occurred since the original bill was introduced.

For An Act To Be Entitled

"AN ACT TO PROVIDE FOR ENVIRONMENTAL EQUITY IN SITING HIGH-IMPACT SOLID WASTE MANAGEMENT FACILITIES BY CREATING A REBUTTABLE PRESUMPTION AGAINST PERMITTING THE CONSTRUCTION OR OPERATION OF ANY SUCH FACILITY WITHIN *TWELVE (12) MILES* OF ANY EXISTING HIGH-IMPACT SOLID WASTE MANAGEMENT FACILITY; TO REPEAL AND SUPERSEDED A.C.A. 8-6-218; AND FOR OTHER PURPOSES."

Subtitle

"AN ACT TO PROVIDE FOR ENVIRONMENTAL EQUITY IN SITING HIGH-IMPACT SOLID WASTE MANAGEMENT FACILITIES."

BE IT ENACTED BY THE GENERAL ASSEMBLY OF THE STATE OF ARKANSAS:

SECTION 1. Legislative Intent.

(a) Through extensive legislation since 1989, the state has made significant strides toward a comprehensive and regionalized approach to solid waste management. The General Assembly recognizes the need to develop viable facilities for managing and disposing of the state's solid waste. This act should be construed as a complement to the state's overall regionalization strategy by encouraging an equitable and efficient dispersal of solid waste management facilities to serve the needs of all citizens.

(b) The General Assembly also acknowledges that, while solid waste management facilities are essential, certain types of facilities impose specific burdens on the host community. National trends indicate a tendency to concentrate high-impact solid waste disposal facilities in lower-income or minority communities. Such facilities may place an onus on the host community without any reciprocal benefits to local residents. The purpose of this act is to prevent communities from becoming the involuntary hosts to a proliferation of high-impact solid waste management facilities.

SECTION 2. Definitions.

The following definitions shall apply for the purpose of this act:

(1)(a) "High-impact solid waste management facility" shall mean, excluding the facilities described in subsection

(1)(b), any solid waste landfill, any solid or *commercial* hazardous waste incinerator, and any *commercial* hazardous waste treatment, storage or disposal facility.

(b) The term "high-impact solid waste management facility" shall not include the following:

(i) Recycling or composting facilities;

(ii) Waste tire management sites;

(iii) Solid waste transfer stations;

(iv) *Solid waste landfills which have applications pending for either increased or new acreage or provisions for additional services of increased capacity;*

(v) A *facility* dedicated solely to the *treatment, storage or* disposal of solid or *hazardous* wastes generated by a private industry where the private industry bears the expense of operating and maintaining the *facility* solely for the disposal of waste generated by the industry or wastes of a similar kind or character; or

(vi) A facility or activity dedicated solely to a response action at a location listed by the state or federal government as a hazardous substance site; or

(vii) *An existing facility operating under interim status of the Federal Resource Conservation and Recovery Act or implementing regulations of the Arkansas Hazardous Waste Management Act or the Arkansas Hazardous Waste Management Code.*

(viii) *Expansion of existing Resource Conservation and Recovery Act or Arkansas Hazardous Waste Management Act hazardous waste facilities, either through increased acreage or provision for additional services or increased capacity.*

(2) "Permitting" means any governmental authorization to proceed with construction of operation of a facility or activity required by either state law or local ordinance.

(3) "Host community" means all governmental units possessing zoning authority encompassed within a *twelve (12)* mile radius of the site of a proposed high-impact solid waste management facility.

(4) "Solid waste" has the same meaning as set out in Arkansas Code 8-6-702 (12), provided however that this definition does not include "hazardous waste" as defined in subsection (5).

(5) "Hazardous waste" has the same meaning as set out in Arkansas Code 8-7-203 (6).

(6) "Hazardous substance sites" has the same meaning as set out in Arkansas Code 8-7-503 (12).

SECTION 3. (a) There shall be a rebuttable presumption against permitting the construction or operation of any high-impact solid waste management facility as defined in this act within *twelve (12)* miles of any existing high-impact solid waste management facility. This presumption shall be honored by the Department of Pollution Control and Ecology, the regional or service area solid waste planning board with jurisdiction over the site, and any other governmental entity with permitting or zoning authority concerning any facility.

(b) The subsection (a) presumption can be rebutted if any of the following is shown:

(1) No other suitable site for such a facility is available within the region or service area because of the restraints of geology or any other factors listed ar *[sic; probably "at"]* Arkansas Code 8-6-706(b)(2); or

(2) Incentives have prompted the host community to accept the siting of the facility. Such incentives may include, without limitation:

(A) Increased employment opportunities;

(B) Reasonable host fees *not to exceed the prevailing state average;*

(C) Contributions by the facility to the community infrastructure (e.g. road maintenance, park development, litter control);

(D) Compensation to *adjacent* individual landowners for any assessed decrease in property values; or

(E) Subsidization of community services.

SECTION 4. Department's Permitting Authority.

The department shall not process any application for a permit subject to Section 3 until the affected local and regional authorities have issued definitive findings regarding the criteria set out in Section 3.

SECTION 5. Specific Repealer.

This act repeals and supersedes the provisions of Arkansas Code 8-6-218.

SECTION 6. All provisions of this act of a general and permanent nature are amendatory to the Arkansas Code of 1987 Annotated and the Arkansas Code Revision Commission shall incorporate the same in the Code.

SECTION 7. If any provision of this act or the application thereof to any person or circumstance is held invalid, such

invalidity shall not affect other provisions or applications of
this act which can be given effect without the invalid proposition or application, and to this end the provisions of this
act are declared to be severable.

SECTION 8. All laws and parts of laws in conflict with
this act are hereby repealed.

Basel Convention on the Control of Transboundary Movements of Hazardous Wastes and Their Disposal: Action in the U.S. Congress

*As with other international treaties, the 1989 Basel Convention on the
Control of Transboundary Movements of Hazardous Wastes and Their
Disposal has no effect in the United States until and unless action is
taken by the U.S. Congress to implement its provisions. During the 1st
session of the 102nd Congress, three bills were introduced into the
House of Representatives to amend the Solid Waste Disposal Act to
bring it into compliance with the Basel Convention. The bills were assigned numbers H.R. 2358, H.R. 2398, and H.R. 2580. The first of
these bills is summarized below. None passed the House, and the United
States has (in mid 1996) still not signed the Basel Convention.*

A Bill

To amend the Solid Waste Disposal Act to ensure that any solid
waste exported from the United States to foreign countries is
managed to protect human health and the environment.

*Be it enacted by the Senate and House of Representatives of
the United States of America in Congress assembled,*

Sec. 1. Short Title

This Act may be cited as the "Waste Export Control Act."

Sec. 2. Findings and Purposes

(a) FINDINGS.—The Congress makes the following
findings:

(1) Exports of solid waste from the United States to
foreign countries are increasing. In several reported instances

exported wastes have been disposed of in a manner that would not be permitted in the United States and are creating foreign policy liabilities for the United States. Many proposals for future waste exports are unsafe.

(2) In some cases, exports of solid waste are being undertaken to avoid higher treatment and disposal expenses in the United States which are associated with the cost of complying with environmental regulations in this country and contribute to the trade deficit of the United States.

(3) Uncontrolled exports of solid waste have a detrimental effect on implementation of existing domestic policy, which recognizes source reduction and recycling as the best methods of solid waste management. Exports should only take place after all reasonable efforts to minimize the generation of the solid waste have taken place.

(4) Existing Federal laws do not provide for any review by the United States of the effects of solid waste exported from the United States on the environment of countries to which the waste is sent.

(5) Uncontrolled export of waste threatens our coasts, oceans, and groundwater through unsound dumping practices.

(6) Compliance by the United States with the 1989 Basel Convention on the Control of Transboundary Movements of Hazardous Wastes and Their Disposal will aid in halting the uncontrolled export of waste to foreign countries.

(b) PURPOSE.—The purpose of this Act is to regulate the export of solid waste from the United States to a foreign country by requiring that such exports be conducted in accordance with an international agreement and strict domestic regulation ensuring that the solid waste is managed in a manner which is protective of human health and the environment and which is no less strict than that which would be required by the solid Waste Disposal Act if the waste were managed in the United States. Compliance with this Act also will satisfy the requirements for United States compliance with the terms of the 1989 Basel Convention on the Control of Transboundary Movements of Hazardous Wastes and Their Disposal.

Sec. 3. Waste Export Control

This section places the proposed legislation into its proper sequence in the U.S. Code by adding a new "Subtitle K—Exports of Solid Waste" to the Solid Waste Disposal Act [42 U.S.C. 6971 et seq.]. It also defines some new terms and prescribes certain conditions for exempted exports. The new sections in the Solid Waste Disposal Act are designated as Sections 12001 and 12002. An important part of the bill is the concluding portion of Section 12002, cited below.

"(b) INTERNATIONAL AGREEMENTS.—(1) Any international agreement pursuant to which solid waste covered by this subtitle may be exported from the United States to another country shall at least include each of the following:

"(A) A provision for notifying the government of the receiving country of exports of such solid wastes.

"(B) A provision for obtaining the consent of the government of the receiving country to accept any solid waste shipment.

"(C) A provision for the United States and the receiving country to exchange information on the manner in which any such solid waste exported from the United States will be managed in the receiving country, including provisions for the exchange of information with respect to the specific treatment, storage, and disposal facilities used for such purposes in the receiving country. Such provisions shall include mechanisms to provide the United States with the information necessary to ensure that transportation, treatment, storage, and disposal of the solid waste will be conducted in a manner which is protective of human health and the environment and which is no less strict than that which would be required by this Act if the solid waste were managed in the United States. Such mechanisms, at a minimum, shall provide a means for the United States to gain access to treatment, storage, or disposal facilities used for the management of such solid waste in the receiving country in the event the Administrator determines such access is necessary to fulfill the Administrator's responsibilities under this subtitle.

"(D) A provision for cooperation between the United States and the receiving country on compliance with and enforcement of the agreement.

"(E) A provision for biennial review by the United States and the receiving country of the effectiveness of the agreement.

"(F) A provision for review and revision or suspension of the agreement if either party concludes that solid waste covered by this subtitle is being transported, treated, stored or disposed of in a manner that is not in accordance with the terms of the agreement.

"(G) A provision which prohibits further transport of such solid waste from the country of destination without the written consent of the parties of the agreement.

"(2) Notwithstanding the provisions of paragraph (1), any bilateral agreement concerning shipments of hazardous waste that has been entered into by the United States and that is in force on the date of enactment of the Waste Export Control Act and which remains in force shall be deemed to meet the requirements of this subsection for a period of two years following enactment of this section. Any such agreement shall comply fully with the provisions of paragraph (1) after the expiration of such two-year period.

"(3) The decision of the United States not to enter into an international agreement shall not be reviewable in any court.

The remaining sections of the act deal with the issuing of waste export permits, user fees, and enforcement procedures.

Court Cases

Margaret Bean et al. v. Southwestern Waste Management Corp. et al.

Cases involving environmental inequities have been difficult to pursue partly because of the U.S. Supreme Court's decision in the case of Washington v. Davis *(see discussion later in this section). In that case, the Court ruled that plaintiffs had to prove "discriminatory intent," that is, that an individual, agency, or corporation specifically intended to cause harm to an individual or a group of individuals. The effect of this ruling can be seen in the first suit brought on the basis of disproportionate exposure to environmental hazards,* Bean v. Southwestern. *In this case, plaintiffs attempt to prevent the construction of a solid waste disposal facility in Houston on the grounds that it had a disproportionate environmental impact on the minority community in which it would have been located. The district court refused to grant a temporary injunction, but allowed plaintiffs to proceed to the discovery stage of their suit. Eventually, however, a different judge dismissed the suit completely.*

I. Introduction

On October 26, 1979, plaintiffs filed their complaint and Motion for Temporary Restraining Order and Preliminary Injunction contesting the decision by the Texas Department of Health to grant Permit No. 1193 to defendant Southwestern Waste management to operate a Type I solid waste facility in the East Houston-Dyersdale Road area in Harris County. *[footnotes omitted]* They contend that the decision was, at least in part, motivated by racial discrimination in violation of 42 U.S.C. 1983 and seek an order revoking the permit. The defendants deny the allegations and have moved to dismiss this case on the grounds of abstention, laches, and the absence of state action. They also complain of the failure of the plaintiffs to name the Texas Department of Water Resources as a defendant . . .

VI. The Preliminary Injunction

There are four prerequisites to the granting of a preliminary injunction. The plaintiffs must establish: (1) a substantial likelihood of success on the merits, (2) a substantial threat of irreparable injury, "(3) that the threatened injury to the plaintiff[s] outweighs the threatened harm the injunction may do to defendant[s], and (4) that granting the preliminary injunction will not disservice the public interest." *[citation omitted]*

The court then goes on to say that plaintiffs have satisfied the second of these conditions, but not the first:

The problem is that the plaintiffs have not established a substantial likelihood of success on the merits. The burden on them is to prove discriminatory purpose. *[Citations to Washington v. Davis and Village of Arlington Heights v. Metropolitan Housing Development Corp. omitted here.]* That is, the plaintiffs must show not just that the decision to grant the permit is objectionable or even wrong, but that it is attributable to an intent to discriminate on the basis of race. Statistical proof can rise to the level that it, alone, proves discriminatory intent, as in *Yick Wo v. Hopkins*, 118 U.S. 356, 6 S.Ct. 1064, 30 L.Ed. 220 (1886), and *Gomillion v. Lightfoot*, 364 U.S. 339, 81 S.Ct. 125, 5 L.Ed.2d 110 (1960), or, this Court would conclude, even in situations less extreme than in those two cases, but the data shown here does not rise to that level. Similarly, statistical proof can be sufficiently supplemented by the types of proof outlined in *Arlington Heights, supra*, to establish purposeful discrimination, but the supplemental proof offered here is not sufficient to do that . . .

If this Court were TDH [Texas Department of Health], it might very well have denied this permit. It simply does not make sense to put a solid waste site so close to a high school, particularly one with no air conditioning. Nor does it make sense to put the land site so close to a residential neighborhood. But I am not TDH and for all I know, TDH may regularly approve of solid waste sites located near schools and residential areas, as illogical as that may seem.

It is not my responsibility to decide whether to grant this site a permit. It is my responsibility to decide whether to grant the plaintiffs a preliminary injunction. From the evidence before me, I can say that the plaintiffs have established that the decision to grant the permit was both unfortunate and insensitive. I cannot say that the plaintiffs have established a substantial likelihood of proving that the decision to grant the permit was motivated by purposeful racial discrimination in violation of 42 U.S.C. 1983. This court is obligated, as all Courts are, to follow the precedent of the United States Supreme Court and the evidence adduced thus far *[footnote omitted]* does not meet the magnitude required by *Arlington Heights, supra.*

VIII. Conclusion

At this juncture, the decision of TDH seems to have been insensitive and illogical. Sitting as the hearing examiner for TDH, based upon the evidence adduced, this Court would have denied the permit. But this court has a different role to play, and that is to determine whether the plaintiffs have established a substantial likelihood of proving that TDH's decision to issue the permit was motivated by purposeful discrimination . . . That being so, it is hereby ORDERED,

ADJUDGED, and DECREED that the plaintiff's Motion for a Preliminary Injunction be, and the same is, DENIED. For the reasons stated above, the defendants' Motions to Dismiss are also DENIED.

Source: 482 F.Supp. 673 (1979)

East Bibb Twiggs Neighborhood Association, et al. v. Macon-Bibb County Planning & Zoning Commission, et al.

This case is similar to the preceding case. Area residents brought suit against the county planning and zoning commission, attempting to prevent construction of a landfill in the neighborhood. The Court rejected the plaintiff's suit, following a line of reason very much like that presented in Bean v. Southwestern, *above.*

Order

OWENS, Chief Judge.

This case involves allegations that plaintiffs have been deprived of equal protection of the law by the Macon-Bibb County Planning & Zoning Commission ("Commission"). Specifically, plaintiffs allege that the Commission's decision to allow the creation of a private landfill in census tract No. 133.02 was motivated at least in part by considerations of race. Defendants vigorously contest that allegation. Following extensive discovery by the parties, this court conducted a non-jury trial on October 4-5, 1988. The parties were permitted to supplement that record following the conclusion of the trial. Based upon a thorough examination of the file and careful consideration of both the evidence submitted and arguments offered during the trial, the court now issues the following ruling. . . .

Discussion

The Court begins this section with a review of relevant precedents, including Bean v. Southwestern *and* Village of Arlington Heights v. Metropolitan Housing Development Corp. *It then continues:*

Having considered all of the evidence in light of the above-identified factors, this court is convinced that the Commission's decision to approve the conditional use in question was not motivated by the intent to discriminate against black persons. Regarding the discriminatory impact of the Commission's decision, the court observes the obvious—a decision to approve a landfill in any particular census tract impacts more heavily upon that census tract than upon any other. Since census tract No. 133.02 contains a majority black population equaling roughly sixty percent (60%) of the total population, the decision to approve the landfill in census tract No. 133.02 of necessity impacts greater upon that majority population.

However, the court notes that the only other Commission approved landfill is located within census tract No. 133.01, a census tract containing a majoritywhite population of roughly seventy-six percent (76%) of the total population. This decision by the Commission and the existence of the landfill in a predominantly white census tract tend to undermine the development of a "clean pattern, unexplainable on grounds other than race . . . *"Village of Arlington Heights,"* 429 U.S. at 266, 97 S.Ct. at 564, 50 L.Ed.2d at 465.

Plaintiffs hasten to point out that both census tracts, Nos. 133.01 and 133.02 are located within County Commission district no. 1, a district whose black residents compose roughly seventy percent (70%) of the total population. Based upon the above facts, the court finds that while the Commission's decision to approve the landfill for location in census tract No. 133.02 does of necessity impact to a somewhat larger degree upon the majority population therein, that decision fails to establish a clear pattern of racially motivated decisions. *[footnote omitted]*

Plaintiffs contend that the Commission's decision to locate the landfill in census tract No. 133.02 must be viewed against an historical background of locating undesirable land uses in black neighborhoods. First, the above discussion regarding the two Commission approved landfills rebuts any contention that such activities are always located in direct proximity to majority black areas. Further, the court notes that the Commission did not and indeed may not actively solicit this or any other landfill application. The Commission reacts to applications from private landowners for permission to use their property in a particular manner. The Commissioners observed during the course of these proceedings the necessity for a comprehensive scheme for the management of waste and for the location of landfills. In that such a scheme has yet to be introduced, the Commission is left to consider each request on its individual merits. In such a situation, this court finds it difficult to understand plaintiffs' contentions that this Commission's decision to approve a landowner's application for a private landfill is part of any pattern to place "undesirable uses" in black neighborhoods. Second, a considerable portion of plaintiffs' evidence focused upon governmental decisions made by agencies other than the planning and zoning commission, evidence which sheds little if any light upon the alleged discriminatory intent of the Commission.

Finally, regarding the historical background of the Commission's decision, plaintiffs have submitted numerous exhibits consisting of newspaper articles reflecting various zoning decisions made by the Commission. The court has read each article, and it is unable to discern a series of official actions taken by the Commission for invidious purposes. *See Village of Arlington Heights*, 429 U.S. at 267, 97 S.Ct. at 564, 50 L.Ed.@d at 465. Of the more recent articles *[footnote omitted]*, the court notes that in many instances matters under consideration by the Commission attracted widespread attention and vocal opposition. The commission oft times was responsive to the opposition and refused to permit the particular development under consideration, while on other occasions the Commission permitted the development to proceed in the face of opposition. Neither the articles nor the evidence presented during trial provides factual support for a determination of the underlying motivations, if any, of the Commission in making decisions. In short, plaintiffs' evidence does not establish a background of discrimination in the Commission's decisions. . . .

The voluminous transcript of the hearings before and the deliberations by the Commission portray the Commissioners as concerned citizens and effective public servants. At no time does it appear to this court that the Commission abdicated its responsibility either to the public at large, to the particular concerned citizens or to the petitioners. Rather, it appears to this court that the Commission carefully and thoughtfully addressed a serious problem and that it made a decision based upon the merits and not upon any improper racial animus.

For all the foregoing reasons, this court determines that

plaintiffs have not been deprived of equal protection of the law. Judgement [sic], therefore, shall be entered for defendants.

SO ORDERED.

Source: 706 F.Supp. 880 (M.D.Ga 1989)

R.I.S.E., Inc., et al. v. Robert A. Kay, Jr., et al.

This case is similar to the two that precede it with one important exception: The court in this case readily acknowledged that the county's policy of landfill siting over the previous two decades had had a disproportionate impact on racial minority communities. This finding was not sufficient, however, to satisfy the Washington v. Davis *standard of discriminatory intent and the court ruled in favor of the defendants.*

The first two-thirds of the court's decision is devoted to a review of the history and other facts in this case. District Judge Richard L. Williams then states his

Conclusions of Law

1. In *Village of Arlington Heights v. Metropolitan Housing Development Corporation*, the U.S. Supreme Court identified the following factors to be considered in determining whether an action was motivated by intentional race discrimination: 1) the effect of the official action; 2) the historical background of the decision; 3) the specific sequence of events leading up to the challenged decision; 4) departures from normal procedures; 5) departures from normal substantive criteria; and 6) the administrative history of the decision. 429 U.S. 252, 266-68, 97 S.Ct. 555, 563-65; 50 L.Ed.2d. 450 (1977).

2. The placement of landfills in King and Queen County from 1969 to the present has had a disproportionate impact on black residents.

3. However, official action will not be held unconstitutional solely because it results in a racially disproportionate impact. Such action violates the Fourteenth Amendment's Equal Protection Clause only if it is *intentionally* discriminatory. [citations omitted]

4. The impact of an official action—in this case, the historical placement of landfills in predominantly black

communities—provides "an important starting point" for the determination of whether official action was motivated by discriminatory intent. *[citation omitted]*

5. However, the plaintiffs have not provided any evidence that satisfies the remainder of the discriminatory purpose equation set forth in *Arlington Heights*. Careful examination of the administrative steps taken by the Board of Supervisors to negotiate the purchase of the Piedmont Tract and authorize its use as a landfill site reveals nothing unusual or suspicious. To the contrary, the Board appears to have balanced the economic, environmental, and cultural needs of the County in a responsible and conscientious manner.

6. The Board's decision to undertake private negotiations with the Chesapeake Corporation in the hope of reaching an agreement to operate a joint venture landfill was perfectly reasonable in light of the county's financial constraints.

7. Once this deal fell through, the Board was understandably drawn to the Piedmont Tract because the site had already been tested and found environmentally suitable for the purpose of landfill development.

8. The Board responded to the concerns and suggestions of citizens opposed to the proposed regional landfill by establishing a citizens' advisory group, evaluating the suitability of the alternative site recommended by the Concerned Citizens' Steering Committee, and discussing with landfill contractor BFI such means of minimizing the impact of the landfill on the Second Mt. Olive Church as vegetative buffers and improved access roads.

9. Both the King Land landfill and the proposed landfill spawned "Not In My Backyard" movements. The Board's opposition to the King Land landfill and its approval of the proposed landfill was based not on the racial composition of the respective neighborhoods in which the landfills are located but on the relative environmental suitability of the sites.

10. At worst, the Supervisors appear to have been more concerned about the economic and legal plight of the county as a whole than the sentiments of residents who opposed the placement of the landfill in their neighborhood. However, the Equal Protection Clause does not impose an affirmative duty to equalize the impact of official decisions on different

racial groups. Rather, it merely prohibits government officials from intentionally discriminating on the basis of race. The plaintiffs have not provided sufficient evidence to meet this legal standard. Judgment is therefore entered for the defendants.

It is so ORDERED.

Source: 768 F.Supp. 1144 (E.D.Va.199)

Washington, Mayor of Washington, D.C., et al. v. Davis et al.

Washington v. Davis *has played an important role in court cases involving environmental equity—and is likely to continue to play such a role—because of the standard of "discriminatory intent" that it establishes. This term means that a plaintiff, such as a plaintiff in an environmental case, must be able to prove that harmful actions taken by an individual, a group, or a corporation were intended to cause harm to the plaintiff and not that the harm occurred as an unexpected by-product of the action. The portion of* Washington v. Davis *in which this position is set forth follows.*

Because the Court of Appeals erroneously applied the legal standards applicable to Title VII [of the Civil Rights Act of 1964] cases in resolving the constitutional issue before it, we reverse its judgment in respondents' favor . . .

As the court of Appeals understood Title VII, employees or applicants proceeding under it need not concern themselves with the employer's possibly discriminatory purpose but instead may focus solely on the racially differential impact of the challenged hiring or promotion practices. This is not the constitutional rule. We have never held that the constitutional standard for adjudicating claims of invidious racial discrimination is identical to the standards applicable under Title VII, and we decline to do so today.

The central purpose of the Equal Protection Clause of the Fourteenth Amendment is the prevention of official conduct discriminating on the basis of race. It is also true that the Due Process Clause of the Fifth Amendment contains an equal protection component prohibiting the United States from invidiously discriminating between individuals or groups. *[citation omitted]* But our cases have not embraced the proposition that a law or other official act, without regard to

whether it reflects a racially discriminatory purpose, is unconstitutional *solely* because it has a racially disproportionate impact.

The Court then reviews a number of cases in which statistical data suggesting *that discrimination has occurred has not been accepted as proof
that an intent to discriminate had existed.*

Necessarily, an invidious discriminatory purpose may
often be inferred from the totality of the relevant facts, including the fact, if it is true, that the law bears more heavily
on one race than another. It is also not infrequently true that
the discriminatory impact—in the jury cases for example,
the total or seriously disproportionate exclusion of Negroes
from jury venires—may for all practical purposes demonstrate unconstitutionality because in various circumstances
the discrimination is very difficult to explain on nonracial
grounds. Nevertheless, we have not held that a law, neutral
on its face and serving ends otherwise within the power of
government to pursue, is invalid under the Equal Protection
Clause simply because it may affect a greater proportion of
one race than of another. Disproportionate impact is not irrelevant, but it is not the sole touchstone of an invidious
racial discrimination forbidden by the Constitution. Stand-
ing alone, it does not trigger the rule, *McLaughlin v. Florida*,
379 U.S. 184 (1964), that racial classifications are to be subjected to the strictest scrutiny and are justifiable only by the
weightiest of considerations . . .

Under Title VII, Congress provided that when hiring
and promotion practices disqualifying substantially disproportionate numbers of blacks are challenged, discriminatory
purpose need not be proved, and that it is an insufficient response to demonstrate some rational basis for the challenged practices. It is necessary, in addition, that they be
"validated" in terms of job performance in any one of several ways, perhaps by ascertaining the minimum skill, ability, or potential necessary for the position at issue and
determining whether the qualifying tests are appropriate for
the selection of qualified applicants for the job in question.
However this process proceeds, it involves a more probing
judicial review of and less deference to, the seemingly reasonable acts of administrators and executives than is appropriate under the constitution where special racial impact,

without discriminatory purpose, is claimed. We are not disposed to adopt this more rigorous standard for the purposes of applying the Fifth and the Fourteenth Amendments in cases such as this.

A rule that a statute designed to serve neutral ends is nevertheless invalid, absent compelling justification, if in practice it benefits or burdens one race more than another would be far reaching and would raise serious questions about, and perhaps invalidate a whole range of tax, welfare, public service, regulatory, and licensing statutes that may be more burdensome to the poor and to the average black than to the more affluent white.

Given that rule, such consequences would perhaps be likely to follow. However, in our view, extension of the rule beyond those areas where it is already applicable by reason of statute, such as in the field of public employment, should await legislative prescription.

As we have indicated, it was error to direct summary judgment of respondents based on the Fifth Amendment.
Source: 426 U.S. 229.

Village of Arlington Heights et al. v. Metropolitan Housing Development Corp. et al.

The "discriminatory intent" rule established in Washington v. Davis *(see above) was reaffirmed only a year later in this case. The case arose because the village of Arlington Heights refused to grant a building permit to Metropolitan Housing for the purpose of constructing racially integrated low- and moderate-income housing. The Village Planning Commission claimed that such a development would damage property values in the village, an argument that the District Court accepted. The Court of Appeals overturned the District Court's decision, however, arguing that the "ultimate effect" of the denial was racially discriminatory since blacks would be prevented from having equal access to public housing. Thus, the Court vacated the District Court's decision even though discriminatory intent was not proved. Upon appeal, however, the U.S. Supreme Court, in turn, overturned the Court of Appeals decision, reiterating that discriminatory intent was the standard that had to be applied in this instance. Selections from Justice Lewis F. Powell's decision follow.*

Respondent Ransom, a Negro, works at the Honeywell factory in Arlington Heights and lives approximately 20 miles away in Evanston in a 5-room house with his mother and his son. The complaint alleged that he seeks and would qualify for the housing MHDC wants to build in Arlington Heights. Ransom testified at trial that if Lincoln Green were built he would probably move there, since it is closer to his job.

The injury Ransom asserts is that his quest for housing nearer his employment has been thwarted by official action that is racially discriminatory. If a court grants the relief he seeks, there is at least a "substantial probability," *Warth v. Seldin, supra,* at 504, that the Lincoln Green project will materialize, affording Ransom the housing opportunity he desires in Arlington Heights . . .

III. Our decision last Term in *Washington v. Davis,* 426 U.S. 229 (1976), made it clear that official action will not be held unconstitutional solely because it results in a racially disproportionate impact. "Disproportionate impact is not irrelevant, but it is not the sole touchstone of an invidious racial discrimination." *Id.,* at 242. Proof of racially discriminatory intent or purpose is required to show a violation of the Equal Protection Clause . . .

Davis does not require a plaintiff to prove that the challenged action rested solely on racially discriminatory purposes. Rarely can it be said that a legislature or administrative body operating under a broad mandate made a decision motivated solely by a single concern, or even that a particular purpose was the "dominant" or "primary" one. *[footnote omitted]* In fact, it is because legislators and administrators are properly concerned with balancing numerous competing considerations that courts refrain from reviewing the merits of their decisions, absent a showing of arbitrariness or irrationality. But racial discrimination is not just another competing consideration. When there is proof that a discriminatory purpose has been a motivating factor in the decision, this judicial deference is no longer justified.

Determining whether invidious discriminatory purpose was a motivating factor demands a sensitive inquiry into such circumstantial and direct evidence of intent as may be available.

The Court then outlines a variety of means by which "invidious discriminatory purpose" might be detected. For example:

Sometimes a clear pattern, unexplainable on grounds other than race, emerges from the effect of the state action even when the governing legislation appears neutral on its face. *[Examples of past Court decisions are cited here.]* The evidentiary inquiry is then relatively easy. But such cases are rare. Absent a pattern as stark as that in *Gomillion* or *Yick Wo*, impact alone is not determinative, and the Court must look to other evidence.

The Court then provides another example:

The historical background of the decision is one evidentiary source, particularly if it reveals a series of official actions taken for invidious purposes. *[Some historical examples are then cited.]* The specific sequence of events leading up to the challenged decision also may shed some light on the decisionmaker's purpose. *[Citations omitted]* For example, if the property involved here always had been zoned R-5 but suddenly was changed to R-3 when the town learned of MHDC's plans to erect integrated housing, we would have a far different case. Departures from the normal procedural sequence also might afford evidence that improper procedures are playing a role. Substantive departures too may be relevant, particularly if the factors usually considered important by the decisionmaker strongly favor a decision contrary to the one reached.

In this case, the Court decides that none of these conditions exist and that there is, therefore, no positive proof of discriminatory intent. On that basis, it reverses the Court of Appeals' decision.

Source: 429 U.S. 252 (1977)

Harrisburg Coalition Against Ruining the Environment v. Volpe

One of the earliest cases dealing with environmental justice issues was that of the Harrisburg Coalition Against Ruining the Environment v. John A. Volpe, *then Secretary of Transportation of the United States. The case involved a number of different issues, one of which was that the construction of highways through a park would prevent black citizens of Harrisburg, Pennsylvania, from having equal access to housing and recreational facilities. In the following decision, Judge Nealon of the*

U.S. District Court for the Middle District of Pennsylvania begins by outlining the case before the court.

This class action arises from the planned construction of two major highways through a public park in the city of Harrisburg, Pennsylvania. Plaintiffs are a community group, the Harrisburg Coalition Against Ruining the Environment, several students and faculty members of the Harrisburg Area Community College (hereinafter HAC), and certain black residents of the Uptown area of Harrisburg. They seek to permanently enjoin the Federal, State and City Governments, as well as the State contractor, from constructing Interstate Route 81 (hereinafter I-81) and the Harrisburg River Relief route through Wildwood Park in the northern section of Harrisburg. . . .

Judge Nealon then points out that seven different issues are raised in the action. The one of most interest here is the third:

Seven issues are raised in this action: . . . (3) whether the construction of I-81 and the River Relief Route through Wildwood Park is a denial to black residents of equal opportunities to housing and recreation in violation of the Fourteenth Amendment and the Civil Rights Acts of 1871 and 1964. . . .

Judge Nealon's opinion on this issue is as follows:

III. Civil Rights Violations

It appears that with respect to the specific Civil Rights claims of plaintiffs, only the denial of equal recreational opportunities is pursued. Indeed, the evidence on the denial of equal housing opportunities is patently insufficient.

The plaintiffs' contention on the recreational opportunity issue can be summarized as follows: by agreeing to give large portions of Wildwood Park to the Commonwealth of Pennsylvania for I-81 and the River Relief route and by failing to maintain Wildwood Park, Harrisburg was partly motivated by an awareness that the predominant use of the Park was by black citizens of the City. The chief evidence advanced by plaintiffs is that the Harrisburg Director of Public Works was alleged to have stated on March 18, 1971, during a meeting on the highway projects in question, in the office of the Governor of the Commonwealth of Pennsylvania, that Wildwood Park was allowed to deteriorate by past City administrators because black people started to use the Park and white people began to go elsewhere for their recreation. When the Director of Public Works testified, he did not deny making the statement attributed to him or that he was representing the Mayor of Harrisburg at the meeting, but he did explain that he prefaced his remarks with a comment that he had been told that this was so. A city resident since only 1959, he could not relate when or by whom he had been told. It was obvious to me that his indictment of past City administrations was without a scintilla of personal knowledge or factual support. This testimony is therefore not accepted.

Furthermore, the credible testimony persuades me that Wildwood Park was used by white residents and black residents of Harrisburg on an equal basis during the many years that Wildwood Park experienced its decline and that certain City officials had made sincere efforts to obtain State and City funds to upgrade Wildwood Park. The deterioration of Wildwood Park is more sensibly attributed to its problems of access, shortage of usable flat land since the inception of HAC, the siltation of the lake, problems of security, and lack of funds for necessary maintenance and improvement, rather than to racial discrimination.

Finally, there is simply insufficient evidence to support the plaintiffs' allegation that the City was improperly motivated in allowing Pennsylvania to use portions of Wildwood Park for the River Relief Route and I-81. Likewise, there is no evidence in this case of any of the kinds of discriminatory results in the administration of municipal services as there were in *Hawkins v. Town of Shaw*, 437 F.2d 1286 (5th Cir. 1971).

Accordingly, I am unable to find that the Mayor or the City of Harrisburg were officially or unofficially motivated in whole or in part by racial reasons in their dealings with Wildwood Park. The same is true of the other defendants. The Civil Rights claims of plaintiffs will therefore be dismissed.

Source: 330 F.Supp. 918 (1971)

El Pueblo Para el Aire y Agua Limpio
v. Chemical Waste Management

*Title VIII of the Civil Rights Act of 1968 makes it illegal "[t]o discrim-
inate against any person in the . . . sale or rental of a dwelling, or in the
provision of services or facilities in connection therewith, because of
race, color, religion, sex, familial status or national origin." Some environmental justice activists have suggested that this section of the law
might be used to prosecute cases in which environmental hazards appear to be—or potentially may be—disproportionately distributed on
the basis of race, color, economic status, or some other criterion. An interesting test of that legal approach is the case named above.*

*In this case, the world's largest waste disposal company, Chemical
Waste Management (CWM), had petitioned the Board of Supervisors of
King County, California, to permit expansion of their waste disposal
site near Kettleman City to allow burning of toxic wastes. The new incinerator was designed to burn 200,000,000 pounds of toxic wastes per
year.*

*With the assistance of two public interest groups (California Rural
Legal Assistance, Inc., and National Housing Law Project), an unin-
corporated association of residents living near Kettleman City, El
Pueblo Para el Aire y Agua Limpio (People for Clean Air and Water),
sued Kings County to prevent it from actually granting the conditional
use permit for the incinerator. The plaintiffs argued, among other
things, that the county's decision was discriminatory because the area
adjacent to the incinerator is low-income, 95 percent Hispanic; because
the siting reflected a similar pattern in CWM's frequent decisions to locate hazardous waste sites in low-income communities of people of
color; and because the county and CWM refused to translate documents
dealing with the issue into Spanish even though nearly 40 percent of the
residents speak only Spanish. The Court made no mention in its decision of the issue of environmental racism and spoke only briefly about
the problem of language. Some important portions of that decision are
as follows.*

The petition for a writ on mandate in the above-entitled proceeding came on regularly for hearing on October 1, 1991,
before the honorable Jeffrey L. Gunther, Judge of the Superior Court [of the State of California in and for the County of
Sacramento]. . .The Court now rules

In this mandate proceeding, petitions challenge a decision of the Kings County Board of Supervisors ("board")

granting a conditional use permit for the construction and operation of a hazardous waste incinerator by Chemical Waste Management, Inc. ("CWM") at CWM's existing hazardous waste treatment, storage and disposal facility in the Kettleman Hills area of southwest Kings County. The board's decision affirmed determinations by the Kings County Planning commission that (1) the environmental impact report prepared on the incinerator project adequately complied with the requirements of the California Environmental Quality Act ("CEQA"; Pub. Resources Code 21000 et seq.) and (2) the incinerator project was consistent with the Kings County General Plan and Zoning Ordinance. Petitions contend that these determinations, and the board's grant of a conditional use permit based on the determinations, are invalid.

Noncompliance with CEQA

For each of the reasons specific below, the Court finds that the Final Subsequent Environmental Impact Report ("FSEIR") on CWM's proposed incineration project was inadequate as an information document under CEQA.

The Court then proceeds to discuss a number of technical areas in which the FSEIR was inadequate, including "Analysis of Air Quality Impacts and Mitigation;" "Analysis of Agricultural Impacts;" "Analysis of Cumulative Air Quality Impacts;" and "Analysis of Project Alternatives." It then alludes briefly to the problem of language.

Public Participation and Access
The Court finds that the strong emphasis in CEQA on environmental decisionmaking by public officials which involves and informs members of the public would have justified the Spanish translation of an extended summary of the FSEIR, public meeting notices, and publish hearing testimony in this case. The residents of Kettleman City, almost 40 percent of whom were monolingual in Spanish, expressed continuous and strong interest in participating in the CEQA review process for the incinerator project at the CWM's Kettleman Hills Facility, just four miles from their homes. Their meaningful involvement in the CEQA review process was effectively precluded by the absence of the Spanish translation.

The Court, however, does not find that the FSEIR was written in a manner incomprehensible to the interested laypersons among the public. The text of the FSEIR perhaps contained significant amount of technical matter which could have been better placed in appendices, but the text was readable. The inadequacies in the analysis, not the readability of the text, constituted the significant deficiency of the FSEIR.

After additional discussion of the incinerator plan's consistency with the county's general zoning plans and of the Board of Supervisor's impartiality in the matter, the court rules as follows.

For the reasons stated above, the Court grants the petition for a writ of mandate and orders the issuance of a writ compelling respondents to set aside the decision certifying the adequacy of the FSEIR and approving a conditional use permit for the construction and operation of a hazardous waste incinerator at CWM's Kettleman Hills Facility. Petitioners are directed to prepare, serve on all parties, and submit to the Court a proposed judgement [sic] and writ of mandate in accordance with this decision.
DATED: DEC 30 1991

Six months later, the court denied a motion by the county and CWM to have a new trial on this issue.

Source: "Ruling," Case No. 366045, Dept. 14, Superior Court, County of Sacramento (California).

Recommendations and Policy Statements

Recommendations to the Presidential Transition Team for the U.S. Environmental Protection Agency on Environmental Justice Issues

The following set of recommendations was developed by a group of nine organizations interested in the issue of environmental justice, operating under the auspices of the Environmental Justice Project of the Lawyers' Committee for Civil Rights under Law. It was submitted to President-

Elect Bill Clinton's Transition Team on 21 December 1992. The complete document is long, and its essence only is summarized below.

Introduction

The Recommendations begin with a brief background describing the nature of the environmental justice movement.

The environmental justice movement is the confluence of three of America's greatest challenges: the struggle against racism and poverty; the effort to preserve and improve the environment; and the compelling need to shift social institutions from class division and environmental depletion to social unity and global sustainability.

This movement has established and documented environmental racism and challenges the existing environmental protection paradigm that results in disparate impact. Race is the most significant predictor of the location of pollution sources ranging from environmental contamination caused by landfills and incinerators, to radiation, pesticide poisoning, and deleterious air quality. Furthermore, occupational exposures and indoor air pollution exacerbate ambient environmental risks.

Environmental justice is not anchored in a debate about whether or not decision-makers should tinker at the edges of risk-based management. The tenets of environmental justice demand implementation of strategies to eliminate unjust and inequitable effects caused by existing environmental policies.

The mission of the U.S. Environmental Protection Agency must be redefined to address environmental laws, regulations and Agency practices that result in discriminatory outcomes. An environmental justice model must be imposed incorporating a framework of equal justice and equal protection principles to ensure every citizen's right to be free from pollution.

The Transition Paper then proceeds to outline recommendations in three areas: the EPA's institutional focus; its regulatory programs, compliance and enforcement activities; and new policy. The main points within each of these areas are as follows:

RECOMMENDATION: A SHIFT TO PROTECTING AD-
VERSELY AFFECTED COMMUNITIES MUST OCCUR IN
EPA'S INSTITUTIONAL FOCUS.
(1) EPA's Office of General Counsel, In Conjunction With
the Department of Justice and the Department's Civil Rights
Division, Should Issue a Formal Opinion Establishing the
Applicability of Civil Rights Laws and Regulations to Envi-
ronmental Programs, and the new Administration Should
Issue an Executive Order Implementing This Policy . . .

*This section explains that the EPA has traditionally argued that civil
rights considerations should not be included in cases with which it
deals. This recommendation suggests that the agency should reverse its
policy in this area and begin to include civil rights laws and regulations
in deciding on environmental issues.*

(2) EPA Should Reassess Governmental Relationships With
Indigenous Peoples, Adequately Fund and Streamline Programs and Facilitate Self-Determination . . .

*The EPA has traditionally held "conflicting approaches" to its dealings
with indigenous peoples, and should now begin to incorporate such peoples at all levels of decision-making about environmental issues.*

(3) EPA Should Put Priority Attention on Developing Countries . . .

*The EPA should make efforts to promote development in developing
nations within the context of protecting the environment of those
nations.*

(4) EPA Should Be Elevated too [sic] Cabinet Status and the
New Administration Should Support Other Key Legislative
Initiatives . . .

*Change in the EPA's administrative status would "facilitate equitable
implementation of statutory programs."*

RECOMMENDATION: EPA SHOULD SUBSTANTIALLY
REORIENT REGULATORY, COMPLIANCE AND EN-
FORCEMENT PROGRAM PRIORITIES . . .
(1) EPA Should Prioritize Environmental Programs to
Redress Disparate Pollution Impact

The eleven program areas to which greater attention should be paid include:

(i) indigenous peoples; (ii) farmworkers; (iii) radiation exposure; (iv) waste facility siting and cleanup; (v) clean air; (vi) clean water; (vii) drinking water; (viii) urban areas; (ix) free trade and border issues; (x) EPA strategic planning and budget; and (xi) state program implementation . . .

(2) EPA Should Target Research & Development Efforts, Including Restructuring the Focus to Reporting and Data Collection on Affected Populations . . .

More reliable data on groups disproportionately affected by environmental pollution should be collected and disseminated.

(3) EPA Should Target Compliance Inspections And Enforcement To Protect Communities of Color Exposed to Disproportionate Environmental Risks . . .

Greater enforcement should be carried out in areas affected by environmental racism.

RECOMMENDATION: NEW POLICY INITIATIVES MUST BE IMPLEMENTED TO REDRESS DISPROPORTIONATE IMPACT

The report lists 14 distinct points relating to the development of new policy initiatives. Some examples include:

—As a means to rebuild infrastructure in communities and around federal facilities, in conjunction with other agencies, states, and educational institutions, EPA should support creation of environmental jobs, training and education in environmental remediation . . .

—The Administration and EPA must revise cost-benefit analysis guidelines to include intangible costs related to quality of life, health, safety and environmental justice . . .

—To establish credibility in EPA programs, the Agency must reverse its historical resistance of cultural diversity and integration in the workforce. EPA should put employees of color in substantive, decision-making position and heed input . . .

—EPA regulations and program should generally shift

the burden of proof to polluters seeking permits in areas which affect highly exposed or multiple-exposure communities.

Source: *Environmental Protection Agency Cabinet Elevation—Environmental Equity Issues.* Hearing before the Legislation and National Security Subcommittee of the House Committee on Government Operations, 103rd Congress, 1st Session, 28 April 1993, pp. 248–262.

Our Calls to Action

The following approach to dealing with problems of environmental racism was outlined by Pat Bryant, Executive Director of the Gulf Coast Tenants Association in a hearing before Congress. Each of the points listed below was discussed in more detail in Bryant's testimony before Congress.

1. Declare a Moratorium on Siting of Poisoning Facilities in the South . . .

2. End Industrial Pollution and Compensate its Victims . . .

3. Make Cleaning Up the Environment the Top National Priority.

4. Provide New Opportunities for Workers Displaced from Poisoning Industries, and Create a Massive Training and Jobs Program for Them and the Unemployed through the Clean-Up . . .

5. Remove Lead Poisoning from our Environment . . .

6. Restore the Sovereignty of Native Lands and Respect and Implement Treaties with Native Peoples . . .

7. Make the South's Workplaces Safe and Healthy . . .

8. Stop the Poisoning of Farm Production . . .

9. Stop the Destruction of Black Farmers and All Small Farms . . .

10. Launch a Massive Housing Construction Program . . .

11. End Academic Tracking and Create Equitable, Adequately Funded Public Schools for All . . .

12. Provide Education Instead of Prisons . . .

13. Provide Full and Adequate Health Care for All . . .

14. Stop the Drug Trade . . .

15. Place Restrictions on Corporations that Move to Other Countries . . .

Source: *Environmental Justice.* Hearings before the Subcommittee on Civil and Constitutional Rights of the House Committee on the Judiciary, 103rd Congress, 1st Session, 3 and 4 March 1993, pp. 13–14.

Principles of Environmental Justice

The following position statement was adopted at the National People of Color Environmental Leadership Summit, Washington, D.C., October 1992.

We, THE PEOPLE OF COLOR, gathered together at this multinational People of Color Environmental Leadership Summit, to begin to build a national and international movement of all peoples of color to fight the destruction and taking of our lands and communities, do hereby re-establish our spiritual interdependence to the sacredness of our Mother Earth; to respect and celebrate each of our cultures, languages and beliefs about the natural world and our roles in healing ourselves; to insure environmental justice; to promote economic alternatives which would contribute to the development of environmentally safe livelihoods; and to secure our political, economic and cultural liberation that has been denied for over 500 years of colonization and oppression, resulting in the poisoning of our communities and land and the genocide of our peoples, do affirm and adopt these Principles of Environmental Justice.

1. *Environmental justice* affirms the sacredness of Mother Earth, ecological unity and the interdependence of all species, and the right to be free from ecological destruction.

2. *Environmental justice* demands that public policy be based on mutual respect and justice for all peoples, free from any form of discrimination or bias.

3. *Environmental justice* mandates the right to ethical, balanced and responsible uses of land and renewable resources in the interest of a sustainable planet for humans and other living things.

4. *Environmental justice* calls for universal protection from nuclear testing, extraction, production and disposal of toxic/hazardous wastes and poisons and nuclear testing that threaten the fundamental right to clean air, land, water, and food.

5. *Environmental justice* affirms that fundamental right to political, economic, cultural, and environmentally self-determination of all peoples.

6. *Environmental justice* demands the cessation of the production of all toxins, hazardous wastes, and radioactive materials, and that all past and current producers be held

strictly accountable to the people for detoxification and the containment at the point of production.

7. *Environmental justice* demands the right to participate as equal partners at every level of decision-making including needs assessment, planning, implementation, enforcement and evaluation.

8. *Environmental justice* affirms the right of all workers to a safe and healthy work environment, without being forced to choose between an unsafe livelihood and unemployment. It also affirms the right of those who work at home to be free from environmental hazards.

9. *Environmental justice* protects the right of victims of environmental injustice to receive full compensation and reparations for damages as well as quality health care.

10. *Environmental justice* considers governmental acts of environmental injustice a violation of international law, the Universal Declaration On Human Rights, and the United Nations Convention on Genocide.

11. *Environmental justice* must recognize a special legal and natural relationship of Native Peoples to the U.S. government through treaties, agreements, compacts, and covenants affirming sovereignty and self-determination.

12. *Environmental justice* affirms the need for urban and rural ecological policies to clean up and rebuild our cities and rural areas in balance with nature, honoring the cultural integrity of all our communities, and providing fair access for all to the full range of resources.

13. *Environmental justice* calls for the strict enforcement of principles of informed consent, and a halt to the testing of experimental reproductive and medical procedures and vaccinations on people of color.

14. *Environmental justice* opposes the destructive operations of multi-national corporations.

15. *Environmental justice* opposes military occupation, repression and exploitation of lands, peoples and cultures, and other life forms.

16. *Environmental justice* calls for the education of present and future generations which emphasize social and environmental issues, based on our experience and an appreciation of our diverse cultural perspectives.

17. *Environmental justice* requires that we, as individuals, make personal and consumer choices to consume as little of Mother Earth's resources and to produce as little waste as

possible; and make the conscious decision to challenge and reprioritize our lifestyles to insure the health of the natural world for present and future generations.

Comments to and About the EPA Environmental Equity Workgroup

Following the January 1990 "Conference on Race and the Incidence of Environmental Hazards" held at the University of Michigan, a group of social scientists and civil rights leaders formed an informal group called The Michigan Coalition to work together on environmental justice issues. One action taken by the Michigan Coalition was to write to William K. Reilly, Administrator of the U.S. Environmental Protection Agency requesting a meeting to discuss possible EPA action on issues of concern to the Coalition. As a result of that request, Reilly convened an Environmental Equity Workgroup to consider ways in which the EPA could respond to issues of environmental equity. The Workgroup was charged with the following four tasks.

Task One: Review and evaluate the evidence that racial minority and low-income people bear a disproportionate risk burden.

Task Two: Review current EPA programs to identify factors that might give rise to differential risk reduction, and develop approaches to correct such problems.

Task Three: Review EPA risk assessment and risk communication guidelines with respect to race and income-related risks.

Task Four: Review institutional relationships, including outreach to and consultation with racial minority and low-income organizations, to assure that EPA is fulfilling its mission with respect to these populations.

Source: *Environmental Equity: Reducing Risks For All Communities, Volume 1: Workgroup Report To The Administrator*, June 1992, pp. 7–8.

The Workgroup took nearly two years to complete its study and then reported its findings to Reilly in a report dated June 1992. Of particular interest are the comments made by groups and individuals interested in issues of environmental equity to draft versions of the Workgroup report. A selection of those comments is reprinted here to give a flavor both of the nature of the concerns of such groups and individuals as well as their opinion of EPA's record on environmental equity in the past.

1. Comments from Bunyan Bryant, Paul Mohai, Benjamin Chavis, Michel Gelobter, David Hahn-Baker, Charles Lee, and Beverly Wright, members of the Michigan Coalition

Comments on Recommendations
"EPA should expand and improve the level and form with which it communicates with racial, minority, and low-income communities, and should increase efforts to involve them in environmental policy making." Increasing the effort is not enough. We want EPA to involve people of color and low-income people in environmental policy making. We recommend that EPA involve people of color in the implementation of the following recommendations. By involving people of color, who have been at the forefront in the environmental equity movement could enhance communication between the agency and people of color. [The following recommendations do not include the extended discussion of each provided in the original document.]

Recommendation 1
"EPA should increase the priority that it gives to issues of environmental equity . . ."

Recommendation 2
"EPA should establish a research and data collection plan and maintain information which provides an objective basis for assessment of risks by income and race . . ."

Recommendation 3
"The EPA should move toward incorporating consideration of environmental equity into the risk assessment process. In calculating population risk, distribution of environmental exposures and risks across the population should be estimated, where relevant. In some cases it may be important to know whether there are any particular population groups at disproportionately high risk . . ."

Recommendation 4
EPA should identify and target opportunities to reduce high concentrations of risk to different population groups, employing approaches developed for geographic targeting . . .

Recommendation 5
"EPA should, where appropriate, selectively assess and consider the distribution of projected risk reduction in major rulemakings and agency initiatives . . ."

Recommendation 6

"EPA should review and selectively revise its permit, grant, monitoring and enforcement procedures to address high concentrations of risk in racial minority and low-income communities. Since state and local governments have primary authority for many environmental programs, EPA should emphasize its concerns about environmental equity to them . . ."

Recommendation 7

"EPA should expand and improve the level and forms with which it communicates with racial minority and low-income communities and should increase efforts to involve them in environmental policy-making . . ."

Recommendation 8

"EPA should establish mechanisms to ensure that environmental equity concerns are incorporated in its long-term planning and operation . . ."

One of the most severe critiques of the Workgroup document came from the Southwest Network for Environmental and Economic Justice. In a long and detailed letter to Reilly, Network reviewers highlighted a number of their most serious concerns about the draft report and then provided a list of its own recommendations to the EPA. The concerns are set off as sidebars within the text of the letter and the recommendations are appended at the end of the letter.

2. Comments from the Southwest Network for Environmental and Economic Justice

"Grassroots people of color organizations have been dealing with 'environmental' problems for decades before the term 'environmental equity' was coined."

"EPA cannot begin to address equity problems until it acknowledges their existence, and this document studiously avoids any such acknowledgement."

"The report, incredibly, fails to even mention the farmworker protection regulations."

[*regarding EPA's lead policies:* " . . . EPA itself has been dragging its feet continually since 1980."

"EPA routinely factors politics and power into its major risk management decisions."

" . . . EPA has never asked Congress for a major appropriation for research."

"EPA is one of the worst agencies in the federal government in terms of integration of its workforce."

" . . . not only does EPA not intend to address environmental equity issue [sic] now, it will not even begin to address these issues until 1996."

" . . . EPA memos detail the Agency's intent to treat environmental equity as a "spin-control" PR exercise with no substantive policy changes reflected anywhere in the Agency's operating guidance."

Recommendations

a. The Agency should develop a major EPA Policy which creates a "presumption of equity" in EPA actions and requires an equity impact analysis for major rules, programs, actions, reviews, etc.

b. EPA should integrate Environmental Equity policy into Operating Year Guidance, strategic plans, and Agency Themes.

c. Pollution Prevention: EPA should work with civil rights groups to implement Pollution Prevention in an equitable way.

c. [sic] Outreach and communication: do not continue attempt to co-opt legitimate leaders. Work with us in mutual respect.

d. EPA should develop formal Federal Register requirements for all State & local grant, permit, delegation, and enforcement policy.

e. The Agency should implement oversight of State & local grant, permit, delegation, enforcement, for equitable implementation.

f. The EPA should establish an Advisory Board with representatives from community-based and labor organizations.

g. The EPA should request funding for data needs.

h. EPA should support a General Accounting Office investigation into whether State programs are in fact equitable.

i. Legislation—EPA should support the Conyers bill, Waxman bill, Chavis bill, and others.

j. Cultural diversity and the integration of the workforce—EPA should put people of color employees in substantive decision making positions and list to input. The Agency should open dialogue and encourage participation of employee organizations in developing overall EPA policy.

k. Structure of Environmental Equity Workgroup; The Workgroup should be assured of its independence. Unions & employee organizations must be involved.

l. Relations with other agencies and organizations; The EPA should work with the US Department of Agriculture, DA, and environmental groups to include equitable considerations and civil rights and labor groups in "power brokered" decisions.

m. EPA should develop an ongoing relationship with the Congressional Black Caucus and other groups.

n. EPA should reopen and reject the 1977 decision withholding application of Civil Rights laws to environmental laws and programs.

o. EPA should immediately issue enforceability provisions of the Farmworker Protection Regulations to make the existing regulations enforceable.

p. EPA should apply the findings of the National Academy of Sciences Report on Pesticides and Children to children exposed in farmworker situations. EPA currently pretends either that children do not work in the fields or that children are no more vulnerable than adults.

Source: *Environmental Equity: Reducing Risks For All Communities, Volume 2: Supporting Document*, June 1992, pp. 83–85; 90–97; and 103–104.

Model Environmental Justice Act

The Center for Policy Alternatives, a nonprofit organization working for progressive policies in all 50 states, has developed a Model Environmental Justice Act. The model act was written with the input and co-operation of 30 legislators from 22 states who make up the Environmental Justice Working Group. The model was first published in December 1994 and then updated in May 1995.

TITLE I:
RISK ASSESSMENT: ANALYSIS OF IMPACT ON COMMUNITIES
SECTION 1: STATE-WIDE IDENTIFICATION OF ENVIRONMENTAL HIGH IMPACT AREAS

(a) ASSESSMENT OF HEALTH RISKS—Within twelve months after the date of enactment of this bill, the Director shall assess the degree of risk to human health posed by toxic chemicals in each county (or other appropriate geographic unit). The Director shall determine the appropriate geographic unit to be used for assessment of risks.

(1) The Director shall publish for public comment, not later than (insert deadline), the methods to be used to assess the degree of risk posed by releases of toxic chemicals, as required under this subsection, as well as the basis for the threshold level of risk determined by the Director to be "substantial" pursuant to subsection (b).

(2) The Director shall publish for public comment, not later than (insert deadline), the methods to be used to calculate the total weight of toxic chemicals released in each county (or other appropriate geographic unit).

(3) For each county (or other appropriate geographic unit) the Director shall calculate and compile the total weight of toxic chemicals released into the ambient environment, broken down by releases into each environmental media [sic] (air, water, land) and by each toxic chemical.

(4) In compiling the data described in subsection (a)(1), the Director shall disregard toxic chemicals which are in a contained, controlled environment such as barrels, factories, warehouses, or lined landfills.

(b) DESIGNATION OF ENVIRONMENTAL HIGH IMPACT AREAS—Within (insert deadline) after the date of enactment, and every (_) years thereafter, the Director shall designate any county (or other appropriate geographic unit) as an EHIA if the degree of risk to human health posed by releases of toxic chemicals in that county (or other appropriate geographic unit) meets a threshold level of substantial risk. This threshold level is to be established by the Director. The Director shall publish a list of the counties (or other appropriate geographic units) of the state, which are designated as EHIAs. The Director shall revise and republish this list every 2 years using the most recent data available.

(c) REPORT ON HEALTH IMPACTS—

(1) Within (insert deadline) after the date of enactment of this bill, the Secretary of Health shall issue a report on EHIAs that shall:

(A) document incidences of cancer, birth deformities, infant mortality rates, and respiratory diseases;

(B) compare the incidence of health impacts under subsection (A) in EHIAs with national, state and demographic averages;

(C) assess the health risks posed by releases of toxic chemicals by individual chemical and cumulatively;

(D) determine the levels to which releases of toxic chemicals, individually and cumulatively, must be reduced so that a county (or other appropriate geographic unit) shall no longer be designated as an EHIA;

(E) determine the impact of releases not regulated by law and releases in violation of current law. This report shall be made available for public review.

SECTION 2: COMMUNITY IMPACT STATEMENTS

(a) IN GENERAL—The Director shall promulgate regulations to require the preparation of a community impact statement as part of the permitting process for any new toxic chemical facility and for any expansion of an existing facility.

(b) PUBLIC REVIEW—Each statement shall be made available for public review, following its release to the local community's elected officials.

(c) SELECTION OF INDEPENDENT CONTRACTOR TO PREPARE STATEMENT—The community impact statement shall be prepared by an independent contractor, who must possess certain qualifications to be defined by the Director. The independent contractor shall be selected by the community's chief elected official, following consultation with community members and the permit applicant.

(d) COSTS—There shall be a fee for each permit application for which a community impact statement is required. The fee shall cover the costs of preparing the community impact statement.

(e) INDICATORS—A community impact statement must provide a detailed summary of findings, written in plain language and limiting the use of technical terms. The statement must identify and describe each of the following:

(1) The number and types of jobs to be created for community members;

(2) The safety standards for the treatment and storage of toxic chemicals;

(3) The proximity of schools and residential areas to the proposed location of the facility;

(4) The facility's emergency contingency plans;

(5) The applicant's record of compliance with state and

federal environmental laws, including its records of compliance in other states and of any firms affiliated with the applicant; and

(6) The presence in the affected community of any other existing toxic chemical facilities and hazardous waste sites.

(f) DEADLINE—A community impact statement must be completed by the independent contractor within (insert deadline) from the date on which the application is filed.

SECTION 3: GRANTS FOR COMMUNITY IMPACT STUDIES OF EXISTING FACILITIES

(a) The Director shall establish a program for the purpose of distributing community impact study grants. The community impact study grants shall:

(1) be funded by user fees levied upon operators of toxic chemical facilities;

(2) enable individuals, citizens groups, and local governments to obtain an independent study of the impact of existing toxic chemical facilities in the area which were sited prior to the requirement of community impact statements under Title I Section 2 of this Act;

(3) detail the effects of the facility on the community's economy, environment, and public health.

(b) To receive a community impact study grant an applicant must present evidence that the community experiences significant:

(1) economic difficulties;

(2) environmental hazards; or

(3) public health problems.

TITLE II:

COMMUNITY EMPOWERMENT AND COMMUNITY-BASED REMEDIAL ACTION

SECTION 1: ENVIRONMENTAL HIGH IMPACT AREAS (EHIAs)

(a) SPECIAL PROGRAMS FOR ENVIRONMENTAL HIGH IMPACT AREAS—

The Director shall establish for communities located in Environmental High Impact Areas the following:

(1) A program enabling communities to hire independent experts to conduct both on-site and off-site monitoring of local facilities to ensure federal laws.

(2) Community environmental resource centers located within existing community service facilities and institutions,

staffed by an environmental expert, which shall do the following:

(A) Provide environmental awareness training to citizens;

(B) Provide education to citizens about state and federal "right-to-know" provisions;

(C) Inform citizens of the opportunities to participate and affect governmental decisions regarding the environment; and

(D) Serve as a clearinghouse for environmental information.

(3) An information and referral service which facilitates collaboration between citizens of an affected community and environmental groups, health experts, and legal advisors who are willing to volunteer their services to promote environmental justice.

(b) ASSISTANCE TO COMMUNITY BASED HEALTH CARE PROVIDERS—The State shall provide grants to community based health facilities in EHIAs to enable them to establish special programs to monitor and respond to adverse health effects on the residents of the community.

(c) INSPECTIONS OF FACILITIES—To ensure the facilities with the highest potential for releases of toxic chemicals are operating in compliance with all applicable environmental health and safety laws and applicable permits, the Director shall conduct periodic inspections of all toxic chemical facilities in EHIAs. The frequency of inspections shall be determined by the Director.

(d) REVIEW BOARD—if the report under TITLE I, Section.1 (d) [sic] identifies significant adverse health impacts from exposure to toxic chemicals, a review board consisting of citizen representatives of the affected communities within the EHIA, working with industry representatives, legislators, and the governor shall propose solutions to remedy and prevent such impacts.

(e) MORATORIUM—if a county (or other appropriate geographic unit) is designated an Environmental High Impact Area (EHIA), there shall be a moratorium in that county (or other appropriate geographic unit) on the siting or permitting of any new toxic chemical facility or any expansion of an existing facility. A new facility or an expansion may be sited or permitted in the county (or other appropriate geographic unit) during the moratorium only if:

(1) The appropriate local government determines in accordance with the Director that there is a pressing environmental need for the new facility or expansion; or

(2) The facility establishes to the satisfaction of the community that any releases of toxic chemicals from the facility will not have a negative impact on public health, and commits to maintaining a comprehensive pollution prevention program.

The moratorium shall continue in effect until the Director determines that the county is no longer designated an EHIA. This determination shall be based on a reassessment of the degree of risk to human health posed by releases of toxic chemicals in each county (or other appropriate geographic unit).

SECTION 2: COMMUNITY-BASED SOLUTIONS

(a) TRUST FUND—The State shall establish a trust fund for community-based environmental clean-up, health testing, and health remediation. The trust fund is to be established through "creative funding" mechanisms (e.g., user fees, expansion of community reinvestment acts, etc.)

(b) SPECIAL LOANS PROGRAM—The State shall create a special loans program to provide resources for community-based environmental clean-up, health testing, and health remediation. The loans shall be financed from income earned by the trust fund described in subsection (A).

(1) Citizen groups can obtain loans up to ($__) in order to fund community-wide environmental clean-up, health testing and health remediation activities.

(2) To receive a loan under this program, an applicant must submit a detailed proposal outlining how they [sic] will use the funds and how the clean-up, testing or remediation will be achieved.

(3) Loans will be forgiven upon satisfactory completion of the proposed clean-up, testing or remediation.

SECTION 3: USE OF COMMUNITY IMPACT STATEMENTS FOR NEW FACILITIES—

(a) CONSIDERATION OF PERMIT APPLICATION— When a Community Impact Statement has been prepared under TITLE I, Section 2, the permitting authority must:

(1) Give great weight to the community impact statement when making any final decision regarding the issuance of a permit;

(2) Deny an applicant its permit, if the statement identifies

any current unabated violations of other permits held by the applicant;

(3) Deny the applicant its permit if it is deemed a "bad actor" because of numerous violations in the past. The Director shall determine the number of past violations necessary to be deemed a "bad actor" as well as the length of the time period to be considered in making this determination; and

(3) [sic] Hold a public hearing at which time members of the community where the site would be located can provide public comments on the community impact statement and other issues relating to the permitting of a facility in their community. The authors of the community impact statement, local government officials, and representatives of the proposed facility must participate in the hearing. The statement and comments made at the public hearing shall be part of the record on which the permitting decision is based.

(b) ISSUANCE OF PERMIT—When a community impact statement prepared under TITLE I, Section 2 identifies a likely significant adverse effect on the community where the facility will be located the State shall take actions to mitigate the effects. The State may attempt to mitigate these effects by supporting community programs relating to employment and economic development, including:

(1) job training and placement programs;

(2) community development corporations;

(3) loans for local businesses;

(4) day care centers for low-income working parents; and

(5) adult education programs.

The Director, in consultation with the appropriate state officials, shall specify which adverse impacts are to be considered significant under this subsection.

SECTION 4: USE OF IMPACT STUDIES OF EXISTING FACILITIES

Independent studies of existing facilities prepared under TITLE I, Section 3, may be used:

(a) to facilitate the filing of citizen petitions for a public hearing;

(b) to request investigation by the Director of the need for remedial action; or

(c) to qualify for State assistance for community programs relating to employment and economic development described in TITLE II, Section 3(b).

SECTION 5: "CLAWBACK" AGREEMENTS

The State shall enable communities to enter into "clawback" agreements with the operators of any new toxic chemical facility. If the local government decides to locate a facility in the community because of promises of economic development and increased employment, they may institute a "clawback" agreement. The "clawback" agreement ensures that if a facility does not satisfy these promises, the community will be compensated monetarily.

TITLE III:

SUSTAINABLE SOLUTIONS FOR STRUCTURAL CHANGE

SECTION 1: TOXIC USE REDUCTION PLANS

(a) CREATION OF PLANS—The Director shall require all toxic chemical facilities in the State to create and implement toxic use reduction plans. Each plan shall evaluate the production or processing methods of the facility, identify possible areas for reduction of the amount of toxic material generated, and outline the methods to be used to implement the reductions. These plans shall be filed with the Director and made available to the public.

(b) ANNUAL REPORTING—Each toxic chemical facility that has created and filed a plan shall file annual updates reporting their progress towards implementing the plan and documenting achieved reductions.

(c) TECHNICAL ASSISTANCE—The Director shall establish a program to provide technical assistance to operators of toxic chemical facilities. The program shall provide information on ways to reduce the amount of toxic material generated by toxic chemical facilities, including analyses of possible areas for reduction in current production and processing methods. The program shall also assist the operators of toxic chemical facilities in the preparation and implementation of individual toxic use reduction plans.

(d) EDUCATIONAL ASSISTANCE—The Director shall establish a program to provide educational assistance to citizens to enable them to:

(1) evaluate a toxic chemical facility's toxic use reduction plan;

(2) evaluate a toxic chemical facility's efforts to implement their reduction plan; and

(3) take advantage of existing opportunities to

participate in and affect state regulation of toxic chemical facilities.

SECTION 2: PUBLIC HEARINGS

(a) The Director shall hold (insert appropriate number) public hearings to investigate issues concerning possible inequities and discrimination in state enforcement of environmental laws.

(b) The Director shall establish citizen advisory committees to ensure direct citizen participation in the hearings.

(c) The Director shall file a report with the legislature which summarizes the hearings, evaluates any concerns voiced by the citizens, and recommends remedies for any existing inequities or discrimination in enforcement.

(d) Additional public hearings shall be held if the Director determines that the need is shown. The Director may make this determination by himself or based upon his review of a citizen petition. The Director shall file a report, as described in Section 1(a), whenever an additional hearing occurs.

SECTION 3: PREVENTION OF CLUSTERING OF TOXIC CHEMICAL FACILITIES

(a) GENERAL PROHIBITION—There shall be a prohibition against permitting the construction or operation of any new toxic chemical facility within *a designed number* of miles of any existing facility.

(b) This prohibition can be waived if, based on public comment from the community where the site is located, the local government:

(1) Determines that pressing local environmental needs require a new facility; or

(2) Decides to accept the siting of a new facility in exchange for incentives offered by the operators of the facility to the community. Such incentives may include, but are not limited to:

(A) Increased employment;

(B) Direct payments to the local government;

(C) Contributions by the facility to the community infrastructure;

(D) Compensation to individual landowners for any assessed decrease in property values; or

(E) Subsidization of community services.

(c) Public comment shall be obtained through hearings, town hall meetings, advisory referenda, and any other appropriate mechanisms.

SECTION 4: SPECIAL INSURANCE

The State shall create a program to assist communities and individuals in purchasing special insurance policies to cover the risk of a future decrease in property values attributable to the siting or operation of a toxic chemical facility.

TITLE IV:

RECOMMENDED SUPPLEMENTAL SECTIONS

SECTION 1: AUTHORITY OF DIRECTOR

The Director shall establish appropriate criteria and standards, where needed, to enforce the provisions of this bill. All proposed criteria and standards established under this section shall be open for public comment prior to final implementation.

SECTION 2: DEFINITIONS

1. *Director*—The head of the state agency responsible for enforcing the State's environmental laws.

2. *Toxic chemical facility*—Proposed definitions: (1) Any solid waste landfill, any solid or commercial hazardous waste incinerator, and any commercial hazardous waste treatment storage or disposal facility. (2) Any facility subject to state or federal environmental regulation.

e. *Environmental High Impact Area* (EHIA)—Any county (or other appropriate geographic unit) that meets a threshold level of risk to human health posed by releases of toxic chemicals in that county (or other appropriate geographic unit). The threshold level of risk is to be determined by the Director.

Community—The area where the toxic chemical facility is located as well as the area generally affected by the facility.

Source: *Center for Policy Alternatives: Model Environmental Justice Act*, Revision of May 1995.

Commission on Racial Justice, United Church of Christ

The study of hazardous waste sites conducted in 1987 by the Commission on Racial Justice of the United Church of Christ was one of the landmark events in the early history of the environmental justice movement. As a result of its findings in that study, the commission made a number of recommendations to governmental and nongovernmental bodies as to how the commission thought the bodies should respond to the problems the commission discovered. Those recommendations are reprinted below.

—We urge the President of the United States to issue an executive order mandating federal agencies to consider the impact of current policies and regulations on racial and ethnic communities.

—We urge the formation of an Office of Hazardous Wastes and Racial and Ethnic Affairs by the U.S. Environmental Protection Agency. This office should insure that racial and ethnic concerns regarding hazardous wastes, such as the cleanup of uncontrolled sites, are adequately addressed. In addition, we urge the EPA to establish a National Advisory Council on Racial and Ethnic Concerns.

—We urge state governments to evaluate and make appropriate revisions in their criteria for the siting of new hazardous waste facilities to adequately take into account the racial and socioeconomic characteristics of potential host communities.

—We urge the U.S. Conference of Mayors, the National Conference of Black Mayors and the National League of Cities to convene a national conference to address these issues from a municipal perspective.

—We urge civil rights and political organizations to gear up voter registration campaigns as a means to further empower racial and ethnic communities to effectively respond to hazardous waste issues and to place hazardous wastes in racial and ethnic communities at the top of state and national legislative agendas.

—We urge local communities to initiate education and action programs around racial and ethnic concerns regarding hazardous wastes.

We also call for a series of additional actions. Of paramount importance are further epidemiological and demographic research and the provision of information on hazardous wastes to racial and ethnic communities.

Source: Commission for Racial Justice, United Church of Christ. *Toxic Wastes and Race in the United States: A National Report on the Racial and Socioeconomic Characteristics of Communities with Hazardous Wastes Sites*. New York: Public Data Access, Inc., 1987, pp. xv–xvi. Used by permission.

Directory of Organizations

The number of organizations interested in the issue of environmental justice has mushroomed in the past decade. Some of these organizations have broad-based interests, ranging from social, economic, and political issues faced by nonwhite and/or lower income populations to environmental issues of every type. Other organizations are focused more specifically on problems of environmental racism, environmental equity, and environmental justice on a local, regional, or national level. By far the most important single resource listing these organizations is *People of Color Environmental Groups*, compiled and edited by Dr. Robert D. Bullard. See chapter 6 for a detailed description of this work. *People of Color Environmental Groups* is an important document because it lists more than 300 individual groups from 40 states, the District of Columbia, Puerto Rico, Canada, and Mexico.

The groups listed in this chapter are those that have indicated a special concern about issues of environmental justice, are primarily national in scope or of unusual regional significance, and have responded to a questionnaire requesting information about the nature of the organization and the work being carried out in the area of environmental justice.

Basic information for groups indicated with an asterisk (*) has been included because of their importance in the environmental justice movement. These groups did not, however, respond to a minimum of five letters, telephone calls, and faxes requesting more detailed information about the organization. The data provided for these groups is believed to be correct, but has not been verified by them.

Agency for Toxic Substances and Disease Registry
1600 Clifton Road, N.E., MS E-28
Atlanta, GA 30333
(404) 639-0700
e-mail: sxc1@atsoaa1.em.cdc.gov

The Agency for Toxic Substances and Disease Registry (ATSDR) is a division of the Public Health Service of the Department of Health and Human Services. Its primary association with the environmental justice movement is through its Minority Health Program (MHP). The program's four major goals concern demographics, health studies and applied research, training and education, and community involvement and risk communication.

The MHP has established specific objectives for accomplishing each of these goals. In the area of demographics, for example, the MHP plans to collect and maintain demographic databases that deal with the siting of hazardous waste sites, identifying communities near hazardous waste sites with predominantly nonwhite populations, and assessing correlations between such demographic patterns and morbidity of affected populations. Health studies and applied research programs are designed to determine the relationship between hazardous substances and health problems in low-income communities and communities of color. In order to meet the training and education goal, the MHP is trying to increase the number of ethnic and nonwhite individuals in professional disciplines related to environmental health issues. To achieve their community involvement and risk communication goal, the MHP is developing educational programs designed to mitigate and prevent health problems in disadvantaged communities and communities of color exposed to environmental hazards.

Americans for Indian Opportunity
681 Juniper Hill Road
Bernalillo, NM 87004

(505) 867-0278
(505) 867-0441 (FAX)

Americans for Indian Opportunity (AIO) was founded in 1970 by
LaDonna Harris in order to serve as a catalyst for finding new
concepts and new opportunities for Indian people. The organization works with tribal governments and peoples to find ways of
dealing with a changing world in ways that are based on traditional tribal values. AIO operates on three fundamental principles. First, the strength of the Indian peoples is in the tribe.
Second, strong tribal governments and communities can have a
positive impact on the world. Third, tribal governments are sov-
ereign units that have the power to determine their own futures.
Among the activities in which AIO is engaged are "Medicine
Pathways for the Future," a national Indian leadership training
project; "Managing Issues through Innovative Tribal Gover-
nance," an issues management and consensus building project;
INDIANet, the first nationally owned and operated Indian
telecommunications computer network; "Intergovernmental Relations," special projects that bring together tribes with federal
and state agencies and departments; and "Evolving Tribal
Economies," a series of studies that demonstrate the contribution
made by tribes to local, regional, and state economies.

Publications: *Messing With Mother Nature Can Be Hazardous To Your
Health*, a report on the environmental health impacts of development on Indian communities.

Asian Pacific Environmental Network
1221 Preservation Park Way, 2nd Floor
Oakland, CA 94612
(510) 834-8920
(510) 834-8926 (FAX)
e-mail: apen@gc.apc.org

The Asian Pacific Environmental Network (APEN) was organized in 1993 in order to provide a vehicle by which Asian Americans and Pacific Islander communities could learn more about
and begin to deal with environmental issues of particular concern to their neighborhoods. The overall mission of APEN is to
"unify and empower the Asian-American and Pacific Islander
communities to achieve multicultural environmental justice."

Activities undertaken by the network are directed at four
major goals: (1) identifying priorities and developing agendas

that will achieve the goal of environmental justice; (2) providing activities that will allow Asian Pacific communities to articulate an environmental justice perspective; (3) building linkages that promote an environmental justice agenda; and (4) identifying resources through which Asian Pacific individuals and groups can redefine environmentalism within their own communities.

Canada Alliance in Solidarity with Native Peoples
39 Spadina Rd.
P. O. Box 574, Station P
Toronto, Ontario M5S 2T1
Canada
(416) 363-4272
(416) 972-6232 (FAX)

The Canada Alliance in Solidarity with Native Peoples (CASNP) was founded in 1960 as the Indian Eskimo Association. It has evolved over the years to reflect changes in the political and social consciousness of its members. Today, the organization works to eradicate non-Native paternalism and promote solidarity between Native and non-Native peoples. Among the organization's seven objectives are efforts to educate non-Native people about indigenous people and their cultures and to counteract misconceptions still persisting in the popular culture; to become a part of the Native people's role as protectors of the land; to provide international support for indigenous peoples everywhere in the world; and to build an information exchange and action network to promote the organization's objectives.

Among CASNP's activities have been fund-raising, education, and public demonstrations around issues such as uranium mining, support for Hopi and Navajo people in Arizona prisons, anti-racism work, and support for local groups such as the Lubicon Cree in northern Alberta, the Innu, and the Mohawks of Kanehsataki. The organization has sponsored social events, educational forums, a "Beyond Racism" conference, coffee houses, concerts, films, Honour Mother Earth Day, and a multicultural Drumfest and Native Rights forum.

Publications: A journal, *Phoenix*; a curriculum resource kit, "All My Relations"; *Resource Reading List*, an annotated bibliography by and about Native peoples; a book, *Indian Giver: A Legacy of North American Native Peoples*; a French language teaching kit for junior level, "Indiens, Inuit, Métis"; and CASNP bulletins on topics such as "Native Women," "Northern Ontario Kit," "Native Land Settlements," "Aboriginal Rights," "Northern Manitoba Flooding," and "Who Owns Canada?"

Center for Policy Alternatives
1875 Connecticut Ave., N.W., Suite 710
Washington, DC 20009
(202) 387-6030
(202) 986-2539 (FAX)
e-mail: cfpa@cap.gwu.edu

The Center for Policy Alternatives (CPA) was established in 1975 as a nonprofit, nonpartisan organization promoting progressive policies in all 50 states. The center provides a network of more than 6,000 state-elected officials and activists whose purpose it is to "develop state public policies that work and then [to] broker successful models from state-to-state to create momentum for national action."

CPA works through a variety of campaigns and projects such as "Women's Voices for the Economy," "Community Capital: Financing the Future," "Mobilizing Participation for Tomorrow's Communities," "The Federalism Initiative," and "Policy for Community Health Providers Project." CPA also works to develop policy models, draft legislation, and convene conferences by working through a network of over 300 state-elected officials called Policy Alternative Leaders. These individuals work together to exchange progressive ideas in all 50 states. CPA also sponsors State Issues Forums, an alliance of more than 60 national organizations that work together to educate state public policy makers about the agenda of nonprofit organizations.

In the field of environmental justice, CPA provides a network through which legislators, activists, and policy leaders can interact on issues in the field. It also provides technical assistance and information in the field of environmental justice. Finally, it has produced a number of publications relating to issues in environmental justice.

Publications: In the field of environmental justice: reports on topics such as "Environmental Priorities and Concerns of State Legislators of Color," "An Ounce of Toxic Pollution Prevention," "Listing of State Legislators of Color," "Toxic Wastes and Race Revisited" (an update of the 1987 United Church of Christ study), "Environmental Justice: Legislation in the States," "Environmental Justice:

Annotated Bibliography," and "Environmental Justice: Model Legislation."

Center for Third World Organizing*
1218 E. 21st Street
Oakland, CA 94609
(510) 533-7583
(510) 533-0923 (FAX)

Established in 1980 to develop and train organizers and leaders to build strong community organizations. Fields of interest: toxics, water pollution, worker safety, housing, environmental justice, and organizing.

Centro de Informacion, Investigacion, y Educacion Social
Cond. El Centro 1, Oficina 1404
Hato Rey, PR 00918
(809) 759-8787/759-8675
(809) 767-6757 (FAX)

A number of communities in Puerto Rico are attempting to deal with the devaluation and demise of traditional and culturally dependent livelihoods that have taken place in recent decades. In such communities, a need has arisen for alternative strategies in areas such as energy, agriculture, fishing, eco-tourism, and reforestation. An organization devoted to assisting communities in such areas is Centro de Informacion, Investigacion, y Educacion Social (CIIES). The primary goal of CIIES is to assist in the development of environmental and community grassroots organization in Puerto Rico by providing research and informational resources, leadership training, and scientific and technical assistance. CIIES operates at all levels of the decisionmaking process, including community discussions, discussions at public hearings, as a liaison between communities and government agencies, and at conferences and public forums. The Center's activities attempt to integrate issues such as toxic pollution, hazardous waste disposal, and land use in providing a foundation for activism based on environmental, social, and economic issues.

Publications: A variety of pamphlets on topics such as the history of environmental struggles in Puerto Rico, toxic pollution, electromagnetic pollution, health problems in public schools, energy, and occupational health.

Children's Environmental Health Network
5900 Hollis St., Suite E
Emeryville, CA 94608
(510) 540-3657
(510) 540-2673 (FAX)

The Children's Environmental Health Network was organized in 1989 to promote the environmental health of children and fetuses. The network claims to be "the only national multidisciplinary and multi-cultural project in the country whose sole purpose is to protect the environmental health of children."

The four primary goals of the network are (1) to promote the development of sound public health and child-focused national policy; (2) to stimulate prevention-oriented research; (3) to educate health professionals, policymakers, and community members in preventive strategies; and (4) to elevate public awareness to the environmental hazards facing children.

The network consists of 18 national organizations, including the American Academy of Pediatrics, the American Academy of Family Physicians, the National Coalition for Hispanic Health and Human Service Organizations, and the Environmental Defense Fund. Some of the activities of the network include a national symposium, "Preventing Child Exposures to Environmental Hazards: Research and Policy Issues"; a national research workshop on pediatric environmental health research (in cooperation with the National Institute for Environmental Health Sciences); 39 scientifically based articles published in the September 1995 issue of *Environmental Health Perspectives*; and the first national train-the-trainers program designed to establish a medical faculty capable of teaching pediatric and family practice residents and graduate nursing students about pediatric environmental health issues.

Citizens Clearinghouse for Hazardous Waste
119 Rowell Court
P. O. Box 6806
Falls Church, VA 22040
(703) 237-2249

Citizens Clearinghouse for Hazardous Waste (CCHW) was founded in 1981 by Lois Gibbs, a housewife in Niagara Falls, New York. Gibbs became interested in the issue of hazardous wastes when she learned that her children were attending a school that had been built on top of a hazardous waste dump. Her campaign on behalf of those living in this "Love Canal" area earned her a national reputation as a fighter against hazardous waste dump sites. Since its founding, CCHW has expanded its agenda to include other environmental issues, such as air and water pollution, solid waste disposal, pesticides, radioactive wastes, and medical wastes.

CCHW collects and disseminates information on this wide variety of environmental issues. It answers questions by telephone and by mail, maintains a library, and produces guidebooks and information packages on environmental issues.

Publications: A quarterly magazine, *Everyone's Backyard*; a monthly magazine, *Environmental Health Monthly*; over 70 guidebooks and information packages on a wide variety of topics.

Citizens Coal Council
110 Maryland Ave., N.E., Room 408
Washington, DC 20002
(202) 544-6210
(202) 544-7164 (FAX)

The Citizens Coal Council was organized in 1989 as a grassroots federation of citizen groups working for social and environmental justice in the area of coal mining. The council consists of about two dozen member groups from 18 states. Included are such groups as the Alabama Environmental Coalition, the People's Action Coalition of Idaho, Kentuckians for the Commonwealth, the Louisiana Environmental Action Network, the Dakota Resource Council, the Buckeye Forest Council, Save Our Cumberland Mountains, the Southern Utah Wilderness Alliance, and the Powder River Basin Resource Council.

The three objectives of the Citizens Coal Council are (1) to protect people, their homes, water, and communities from mining damage; (2) to obtain enforcement of the federal Surface Mining Control and Reclamation Act; and (3) to "help each other win our issues."

Publication: A magazine, *Reporter*.

Citizens' Environmental Coalition
33 Central Avenue
Albany, NY 12210
(518) 462-5527
(518) 465-8349 (FAX)
e-mail: cectoxic@igc.apc.org

Citizens' Environmental Coalition is a group of individuals and groups working on problems of pollution, hazardous wastes, pesticides, and incinerator ash in New York State. The coalition was organized to provide a mechanism by which citizens and groups could share skills, resources, and information about ways of preventing and dealing with toxic, radioactive, and pesticide contamination. Its goal is to eliminate these forms of pollution in homes, schools, workplaces, and the physical environment.

Publications: A quarterly newsletter, *Toxics in Your Community Newsletter*; a large annotated map of Superfund sites titled "Toxic Dumps are Poisoning Our Environment, Health & Economy"; and factsheets on topics such as "The Right to Know About Toxic Chemicals in Your Community," "How to Create a Toxic Plume Map," "At Risk: Transporting Hazardous and Radioactive Materials," and "Managing Medical Wastes."

Citizens for a Better America
P. O. Box 356
Halifax, VA 24558
(804) 476-7757

Citizens for a Better America (CBA) was founded by Cora Tucker in 1975 with the purpose of making "America a better place for all." "The treatment I received as a young child in Halifax County," Ms. Tucker has written, "made me want to make some changes. Growing up black in Southside Virginia in the 1940s and '50s was pure hell."

CBA has been involved in a number of activities designed to improve the life of blacks in the Halifax area. These include lobbying, research, organizing, educating, and carrying out direct action in the areas of toxic materials, air and water pollution, waste disposal, facility siting, recycling, parks and recreation, housing, community organizing, and environmental justice.

Publications: A scrapbook describing articles, letters, speeches, and other activities carried out by CBA, "20 Years of Hard Work"; a booklet describing the organization's work, "12th Citizenship Day of Prayer."

Commission for Racial Justice
United Church of Christ
700 Prospect Avenue
Cleveland, OH 44115-1110
(216) 736-2100
(216) 736-2171 (FAX)

(Most environmental justice activities are handled through the following office in New York City.)
475 Riverside Drive, Room 1948
New York, NY 10115
(212) 870-2077
(212) 870-2162 (FAX)
(Issues of a rural character are handled by the following office in Enfield, North Carolina.)
Office for Rural Racial Justice
P. O. Box 187
Enfield, NC 27823

The Commission on Racial Justice was formed by the United Church of Christ in 1963 in response to bombings of Birmingham churches, the assassination of Medgar Evers, and other events tending to heighten racial tensions during the early years of the decade. The commission states that it is "committed to working tirelessly for racial justice for racial and ethnic people—African Americans, Hispanic Americans, Native Americans and Asian Americans and for reconciliation of all people." Specific programs developed by the commission have focused on issues such as child abuse, teenage pregnancy, capital punishment, voter registration, health care, family life, penal reform, and environmental justice. The commission's report on "Toxic Wastes and Race in the United States" is widely recognized as one of the most important documents in the early history of the environmental justice movement. The commission maintains regional offices in Washington, D.C. (Office for National and Urban Racial Justice), Enfield, North Carolina (Office of Constituency Development and Rural Racial Justice), and New York (Office of Ecumenical Racial Justice) in addition to its national office in Cleveland.

Publications: An informational pamphlet, "The Commission for Racial Justice Of the United Church of Christ," "Pastoral Letter on Contemporary Racism and the Role of the Church," and a variety of pamphlets and other publications, such as "Racism and Public Education," "Teenage Pregnancy Prevention: A Rites of Passage Manual," "The Black Family: An Afrocentric Perspective," and "The Black Mentally Retarded Offender."

Deep South Center for Environmental Justice
Xavier University of Louisiana
7325 Palmetto Street, Box 45B
New Orleans, LA 70125

(504) 483-7340
(504) 488-7977 (FAX)

The Deep South Center for Environmental Justice (DSCEJ) was founded in 1993 to provide a means by which universities and communities could work together to deal with issues of environmental justice. The center's location at Xavier University, a historically black university, is fortuitous because of its location near many of the communities in Louisiana most severely affected by environmental hazards.

The center's three key objectives are (1) partnership between universities and communities; (2) interaction between program components; and (3) legacy. The three forms of activity by which the center attempts to accomplish these goals are (1) research and policy analysis and development; (2) community assistance and education; and (3) education at both precollege and college levels.

EcoJustice Project
Center for Religion, Ethics, and Social Policy
Anabel Taylor Hall
Cornell University
Ithaca, NY 14853-1001
(607) 255-5027

The EcoJustice project was a project of the Center for Religion, Ethics, and Social Policy at Cornell University. It was based on a concept that regards ecological wholeness and social-economic justice as necessary and inseparable components of a desirable future. According to the director of the center, the EcoJustice project had "begun some interesting work on environmental racism, but it was not funded." As a result, its staff was laid off, and its quarterly journal, *The Egg*, was discontinued.

EcoNet
Institute for Global Communications
18 De Boom Street
San Francisco, CA 94107
(415) 442-0440
(415) 546-1794 (FAX)
e-mail: support@igc.apc.org

The Institute for Global Communications (IGC) is an international community of activists who use computer networking as a means of bringing about progressive change in the fields of human rights, social justice, environmental sustainability, conflict resolution, women's empowerment, and the equitable treatment of workers. The institute currently reaches over 28,000 activists in 133 countries, with information available to millions more through the Internet.

The five major divisions of the institute include PeaceNet, EcoNet, LaborNet, ConflictNet, and WomensNet. One of the focus areas within EcoNet is Environmental Justice and Racism. Information about this field is available from more than 4,000 EcoNet members, including the American Farmland Trust, Berkeley Ecology Center, EarthNet, Environmental Liaison Center International, Environmental Defense Fund, Friends of the Earth, Greenpeace, National Resources Defense Council, and the Pesticide Action Network.

Publications: Access to hundreds of news services and publications worldwide, including *African American News Service*, *Green Left Weekly*, *Mother Jones*, *Pacific News Service*, and *The Earth Times*. IGC also provides a number of other services, such as teleconferencing, e-mail, consulting and technical advice, electronic mailing lists, and custom on-line services.

Environmental Justice Resource Center
Clark Atlanta University
223 James P. Brawley Drive, S.W.
Atlanta, GA 30314
(404) 880-6910
(404) 880-6909 (FAX)
internet: rbullard@cau.auc.edu

The two major goals of the Environmental Justice Resource Center are "(1) to facilitate the dissemination of scientific, technical, and legal education and information in environmental high-impact communities and among environmental justice groups and other nongovernmental organizations; and (2) to design community-focused environmental justice model demonstrations on environmental technology transfer, training, risk communication, pollution prevention, and public participation with low-income and people of color communities." Some of the center's activities include community outreach to low-income and people of color communities, community-partnership research, maintenance of archives on environmental justice, a speakers clearinghouse, an environmental justice database and resource directory, and a training institute.

Environmental Research Foundation
P. O. Box 5036
Annapolis, MD 21403-7036
(410) 362-1584
(410) 263-8944 (FAX)
e-mail: erf@rachel.clark.net
econet: erf
internet: erf@igc.apc.org

The Environmental Research Foundation provides newsletters, resource guides, resource sheets, fact sheets, and other materials related to issues of environmentalism, in general, and environmental
justice, in particular. Examples include "Research Corporations: A
Guide to Organizations," "Petroleum Report," "Environmental
Whistleblowers: An Endangered Species," "Students' Guide to
Helping the Environment," "Dioxin Package," "Pesticide Regulation Package," and "Incineration Package." The foundation is perhaps best known for its newsletter, *Rachel's Environment & Health
Weekly* (formerly, *Rachel's Hazardous Waste News*).

Publications: See above.

Gulf Coast Tenants Organization*
1866 N. Gayoso Street
New Orleans, LA 70119
(504) 949-4919
(504) 949-0422 (FAX)

Founded in 1982 to train people to become organizers and leaders. Issues of interest: toxics, energy, air and water pollution,
waste disposal, facility siting, pesticides, lead, asbestos, parks
and recreation, worker safety, housing, environmental justice,
community organizing, and economic development.

Harvard Center for Risk Analysis
Harvard School of Public Health
718 Huntington Ave.
Boston, MA 02115
(617) 432-4497
(617) 432-0190 (FAX)
e-mail: mrocha@sph.harvard.edu

The Harvard Center for Risk Analysis (HCRA) was established
within the Harvard School of Public Health in 1989 as a way of
"promoting a reasoned public response to health, safety, and

environmental hazards." The center's goal is to bring to bear on important national issues of these kinds a careful, scientific analysis that can lead to informed public policy decisions about the best way of handling such problems. Some examples of current projects sponsored by the center include Risk-Based Superfund Reform, Characterizing Chemical Risks in Data Sparse Situations, Guidance for Distributional Analysis of Exposures, and Risk Equity. Among the activities supported by the center are research studies of the type cited above, testimony before Congress, development of new methods for dealing with issues of risk analysis, training and education of specialists in the field of risk analysis, the support of visiting scholars, and the publication of books, articles, and other publications dealing with risk analysis.

Publications: Occasional publications, such as *A Historical Perspective on Risk Assessment in the Federal Government, Risk Assessment in the Federal Government*, and *Reform of Risk Regulation: Achieving More Protection at Less Cost*; members of the center have also written a number of articles and books in connection with their research. For more information about these publications, contact Melissa Rocha at the address above.

Highlander Education and Research Center
1959 Highlander Way
New Market, TN 37820
(423) 933-3443
(423) 933-3424 (FAX)

Highlander was founded in 1932 as the Highlander Folk School. The purpose of the center is to work with people who are struggling against oppression by supporting their attempts to take collective action to shape their own destinies. It accomplishes this goal by providing a variety of educational experiences that empower people to take democratic leadership towards fundamental change.

During the six decades of its history, the Highlander Center has focused on such issues as worker education in the 1930s, citizenship schools for southerners of African descent in the 1950s, civil rights schools in the 1960s, and building communication organizations in Appalachia in the 1970s and 1980s.

Among the honors accorded the center have been a two-hour PBS special hosted by Bill Moyers in 1981 and nomination for the Nobel Peace Prize in 1982.

The center's mission statement explains that it "bring[s] people together to learn from each other. By sharing experience, we realize that we are not alone. We face common problems caused by injustice. By affirming our cultural and racial diversity, we overcome differences that divide us. Together we develop the resources for collective action. By connecting communities and groups regionally, we are working to change unjust structures and to build a genuine political and economic democracy."

Among the many resources provided by the center are workshops and educational training sessions conducted at its New Market farm and center, a library and audiovisual resource center, regional workshops conducted by center staff, and its leadership programs for young people and interns.

Publications: A quarterly newsletter, *Highlander Reports*; CDs and tapes, videos, and other special publications. Some examples of titles include *Dismantling the Barriers: Rural Communities, Public Participation, and the Solid Waste Policy Dilemma* (handbook); *Environment and Development in the USA: A Grassroots Report* (manual); *Always Sing My Song, Come All Ye Coalminers and Save the Land and People* (video); *You Got To Move* (video); *Water: You Have To Drink It with a Fork* (working paper); and *Pollution Industries in the South and Appalachia: Economy, Environment, and Politics* (working paper).

Indian Law Resource Center
602 North Ewing Street
Helena, MT 59601
(406) 449-2006
(406) 449-2031 (FAX)
Washington office:
601 E St., S.E.
Washington, DC 20003
(202) 547-2800
(202) 547-2803 (FAX)

Land rights, environmental protection, native sovereignty and self-government, human rights, and law reform are the major issues in which the Indian Law Resource Center is involved. Founded in 1978, the center is a tax-exempt organization under section 501(c)(3) of the Internal Revenue Code. It is funded entirely by grants and contributions from Indian nations and receives no governmental support. The center's principal goal is "the preservation and security of Indian and other Native nations and tribes."

Examples of some of the issues on which the center has worked in recent years include finding a new site on which the Traditional Seminole Nation will be able to conduct its Green Corn Dance Ceremony, protection of the Gros Ventres and Assiniboine Tribes in Montana from the effects of cyanide heap-leach mining adjacent to their reservation, negotiations on the Miskito Coast Protected Area Project in Nicaragua, assistance with the development of the Toledo Maya Homeland Proposal in southern Belize, and activities to address human rights abuses against the Yanomami Indians in Belize.

Publications: Annual reports and a quarterly newsletter, *Indian Rights; Human Rights.*

Institute for Local Self-Reliance
2425 18th Street, NW
Washington, DC 20009-2096
(202) 232-4108
(202) 332-0463 (FAX)
e-mail: ilsr@igc.apc.org

The Institute for Local Self-Reliance was established in 1974 for the purpose of helping grassroots community groups, government leaders, and business entrepreneurs develop and implement environmentally sound economic development strategies. The organization originally focused on the Adams Morgan neighborhood of Washington, D.C., developing solar collectors, rooftop hydroponic gardens, a composting toilet, and a basement sprout operation. It later broadened the scope of its activities to include whole cities and then regions. Its vision has been to show neighborhoods and communities how to "extract the maximum value from their local human, capital and natural resources."

Publications: More than 70 books and monographs on topics such as *Self-Reliant Cities, Creating Local Recycling Markets, In-Depth Studies of Recycling and Composting Programs, Waste Prevention, Recycling and Composting Options: Lessons from 30 U.S. Communities, An Environmental Review of Incineration Technologies,* and *Institute for Local Self-Reliance Youth Planning Manual*; a newsletter, *Self-Reliance Newsletter.*

Labor Institute
853 Broadway, Room 2014
New York, NY 10003

(212) 674-3322
(212) 353-1203 (FAX)

The Labor Institute is a nonprofit education and research organization that provides innovative worker-oriented training programs and materials to unions and community groups around the United States. The staff are members of OCAW (Oil, Chemical, and Atomic Workers) Local 8-149. The institute has carried out a vigorous program of developing materials to inform workers and the general public about the hazards that exist in workplaces and in residential and industrial areas.

Publications: Workbooks and curriculum materials, such as *Tuberculosis in the Workplace, Jobs and the Environment, OCAW-Labor Institute Hazardous Waste Training Workbook*, and *Corporate America and the American Dream: Toward an Economic Agenda for Working People*; videotapes, such as *More than We can Bear: Reproductive Hazards on the Job, Trust in Training*, and *A Price for Every Progress: The Health Hazards of VDTs*.

Labor Occupation Health Program
Center for Occupational and Environmental Health
School of Public Health
University of California at Berkeley
2515 Channing Way
Berkeley, CA 94720-5120
(510) 642-5507
(510) 643-5698 (FAX)

The Labor Occupational Health Program (LOHP) was established in 1974 to provide health and safety training, information, and assistance to unions, workers, joint labor-management groups, and health professionals. Among the services provided by LOHP are training on toxic substances, hazardous wastes, stress, workplace design, and other occupational hazards; technical assistance; videotapes, slides, and print publications; a resource center of health and safety books and other materials; a program of continuing education; and special projects designed to meet the program's objectives.

Publications: A newsletter, *LOHP Monitor*; books on a variety of topics, including *Participatory Methods for Hazardous Waste Trainers, Work and Health in the Latino Community, Occupational Disease among Black Workers, Health and Safety Issues Commonly Faced by Farmworkers, Spanish Health & Safety Training Manual*, and *Chemical Hazards in the Building Trades*; pamphlets, such as "Is Work Making

You Sick?" and "Are You a Hazardous Waste Worker?"; teaching units, such as "Construction Workers Student Guide" and "Welders Student Guide"; and factsheets, such as "How Toxics on the Job Affect Your Health" (in English, Spanish, and Chinese) and "Should Unions Be Concerned about Smoking At Work?"

Madres del Este de Los Angeles
(Mothers of East Los Angeles; MELA)
924 South Mott Street
Los Angeles, CA 90023
(213) 269-9898
(213) 269-2446 (FAX)

Not economically rich—but culturally wealthy, not politically powerful—but socially conscious, not mainstream educated—but armed with the knowledge, commitment, and determination that only a mother can possess. This mission statement describes the philosophy that has driven MELA for more than a decade. The organization was formed on 24 May 1984 with the sole purpose of fighting the construction of a new state prison in the East Los Angeles neighborhood of Boyle Heights. When this campaign was successful, MELA decided not to disband, but to consider its efforts to improve the environmental quality of their neighborhood.

In 1987, for example, the members of MELA worked to prevent the construction of a municipal waste incinerator and an oil pipeline that would have passed 3 feet under a local junior high school. A year later, the organization successfully prevented the City of Vernon from building a toxic waste incinerator and, a year later, stopped the Chem-Clear Plant from installing a toxic waste treatment plant near Huntington Park High School.

In recent years, MELA has moved on to other projects. In 1992, for example, it initiated a low-flush toilet program designed to conserve water that has won recognition from organizations as varied as the Sierra Club to the Mono Lake Committee to Tree People. Revenue obtained from the ultra-low-flush toilet program has enabled MELA to hire 31 people to work on its water conservation program and the immunization program that it operates in cooperation with White Memorial Medical Center.

Migrant Legal Action Program, Inc.
2001 S Street, N.W., Suite 310
Washington, DC 20009
(202) 462-7744

The Migrant Legal Action Program (MLAP) has been working for nearly 25 years to obtain justice for the more than three million migrant farmworkers who harvest the nation's crops. MLAP works at both the national and local level to accomplish its goal. It assists local legal services programs and farmworker organizations by providing community education sessions and advocacy training; answers to written and telephone requests for assistance; periodic mailings on current issues affecting migrants; and a range of publications on migrant issues. At the national level, MLAP is engaged in litigation on migrant problems; acts as an advocate at administrative and legislative hearings; represents farmworkers in administrative proceedings; and provides leadership to and coordination of farmworker groups on national issues.

Publications: A biweekly newsletter, *Field Memo*; numerous other educational materials, manuals, and other publications.

National American Indian
Environmental Illness Foundation
Box 1039
Long Beach, WA 98631
(360) 665-3913
(360) 642-4467 (FAX)

The National American Indian Environmental Illness Foundation was created in 1994 in order to provide information to Native Americans who have been afflicted with chemically induced disorders such as chronic fatigue/immune dysfunction syndrome, multiple chemical sensitivity, environmental illness, Gulf War syndrome, sick building syndrome, and fibromyalgia syndrome. The foundation is compiling a list of physicians and practitioners, both Native American and non-Native American, who are knowledgeable about such disorders. It provides educational materials and a resource list to those who are interested in developing a support circle for patients with such disorders.

Publications: An informational packet on environmentally related illnesses.

National Coalition against the Misuse of Pesticides
701 E Street, S.E.
Washington, DC 20003
(202) 543-5450
(202) 543-4591 (FAX)

The National Coalition against the Misuse of Pesticides was founded in 1981 as "a broad coalition of health, environmental, labor, farm, consumer, and church groups, as well as individuals, who share common concerns about the potential hazards associated with pesticides." Some of the organizations currently represented on the coalition's board of directors include Pesticide Watch (San Francisco), Northwest Coalition for Alternatives to Pesticides (Eugene, Oregon), Kansans for Safe Pest Control, Colette Chuda Environmental Fund (Malibu, California), Farm Labor Organizing Committee, and Agricultural Resources Center (Carrboro, North Carolina).

The threefold focus of the coalition is to evaluate pesticide use and pest management practices, to document the effects of pesticides on human health and the environment, and to help people organize for the adoption of alternatives to the use of pesticides. Information about pesticides is available from the Coalition's Center for Community Pesticide and Alternatives Information.

Publications: Books, such as *A Failure To Protect: The Unnecessary Use of Hazardous Pesticides at Federal Facilities Threatens Human Health and the Environment*; *Unnecessary Risks: The Benefit Side of the Pesticides Risk Benefit Equation*; and *Safety at Home: A Guide to the Hazards of Lawn and Garden Pesticides and Safer Ways to Manage Pests*; packets, such as "Activism," "Alar®," "Chemical Sensitization," "Golf Courses," "Insect Repellents," "Mercury in Paint," "Reproductive Effects," "School IPM," and "Wood Preservatives"; brochures, such as "Pesticides & Your Fruits and Vegetables," "Least Toxic Control of Lawn Pests," and "Estrogenic Pesticides: What You Need to Know & What You Need to Do"; and other publications, such as "The Great American Water Debate," "Citizens Guide to Rights-of-Way Management," "Pesticides and Schools," "Into the Sunlight: Exposing Methyl Bromide's Threat to the Ozone Layer," and "Model Ordinance Book."

National Conference of Black Lawyers
1875 Connecticut Ave., N.W., Suite 400
Washington, DC 20009
(202) 234-9735
(202) 234-9743 (FAX)

One of the many positive results arising out of the civil rights movement of the 1960s was the realization by black attorneys that their skills were important tools in the ongoing battle against racism and for the liberation of African Americans. As a result, a group of black attorneys came together at Capahosic, Virginia, in 1968 to found the National Conference of Black Lawyers (NCBL). That meeting was followed a year later by a second and larger conference held in Chicago at which NCBL was officially established. The guiding principles of the organization are expressed in the Capahosic Declaration of Concern and Commitment, which states that the purpose of NCBL is to create "a permanent and ongoing body of Black lawyers determined to join the Black revolution and committed to taking all steps to assist Black people in attaining the goals to which they are rightly entitled...law, morality and justice."

NCBL's work falls into three major areas: criminal justice; social, economic, and political justice; and international affairs and world peace. In the area of environmental justice, its activities include the development of legal strategies and general support for the grassroots movement seeking environmental justice, which includes opposition to racially discriminatory enforcement of environmental laws; the deliberate targeting of communities of color and the poor for toxic waste; and the exclusion of people of color from leadership in the environmental movement.

Although NCBL maintains a national office in Washington, D.C., much of its work is done through local chapters located in more than 20 regional sites, including Berkeley, Venice, and Oakland, California; Kew Garden, New York; New Orleans; Atlanta; Chicago; St. Thomas, U.S. Virgin Islands; Camden, New Jersey; Jackson, Mississippi; Philadelphia; and Dallas.

National Congress of American Indians
2010 Massachusetts Avenue, N.W., Second Floor
Washington, DC 20036
(202) 466-7767
(202) 466-7797 (FAX)

The National Congress of American Indians is the oldest and largest national Indian tribal government advocacy organization. It was established in 1944 for the purpose of promoting and protecting the sovereignty and treaty rights of American Indian and Alaska Native nations. The congress currently has a membership of over 200 tribal governments.

An example of the work done in the field of environmental justice is the congress' comments on the Draft Waste Management Programmatic Environmental Impact Statement developed by the U.S. Department of Energy (DOE). The congress was particularly critical of DOE's intention to locate 9 of the 17 "major" sites within 50 miles of a federally recognized Indian tribe. It commented at some length on DOE's failure to recognize or deal with "the clearly evidenced disproportionate impacts on the tribes."

National Institute of Environmental Health Sciences
National Institutes of Health
Office of Communications
P. O. Box 12233
Mail Drop WC-03
Research Triangle Park, NC 27709-2233
(919) 541-1402
(919) 541-2242 (FAX)
e-mail: hawkins@NIEHS.NIH.gov

The National Institute of Environmental Health Sciences (NIEHS) was established as a division of the National Institutes of Health in 1966. It serves as the principal federal agency for basic biomedical research and training on the health effects of environmental agents and on preventing, intervening in, and treating diseases and dysfunctions associated with the environment. In recent years, NIEHS has supported a number of activities in the field of environmental justice, including the creation of three Environmental Health Science Research Center Programs in Louisiana, Kentucky, and New York to encourage research on environmentally related health problems of the poor and ethnic minorities; research on the health effects of agricultural chemicals, on the source of blood lead during pregnancy, on the genetic basis of increased susceptibility of African Americans to cigarette smoke with regard to the development of bladder and lung cancer, and on the transfer of blood lead from mothers to fetuses; support for the involvement of poor and minority communities in research on environmental health issues; and sponsorship of various conferences, symposia, and workshops aimed at developing a research agenda for environmental justice.

Publications: Technical reports; informational pamphlets, such as "Lead and Your Health" and "Mi Mundo Bajo Techo."

Native Americans for a Clean Environment
P. O. Box 1671
Tahlequah, OK 74465
(918) 458-4322
(918) 452-0322 (FAX)

Native Americans for a Clean Environment (NACE) was founded in 1985 by a group of 12 Native American and non-Native American people in opposition to the activities going on at the Sequoyah Fuels uranium conversion facility in Carlile, Oklahoma. Carlile is a rural community whose residents are mostly members of the Cherokee tribe. Over the next decade, NACE experienced a number of successes that included stopping the proposed disposal of radioactive wastes by deep well injection, stopping the expansion of radioactive waste storage ponds, intervening in water quality permit hearings at the plant, and working with the Coalition to Stop Food Irradiation to defeat a plan to build a food irradiator in Choctaw County in southeastern Oklahoma. In 1992, Sequoyah Fuels announced permanent closure of its facility in Carlile.

NACE also works with other Native American tribal groups and governments, providing assistance in organizing and educating people about environmental problems facing today's Indian peoples. Thus far, NACE has worked with peoples from the Acoma, Apache, Cherokee, Cheyenne, Choctaw, Cree, Dakota, Diné, Hopi, Kiowa, Laguna, Lakota, Los Coyotes, Meti, Mohawk, Muscogee, Paiute, Shawnee, Soshone, Ute, Yakima, and Zuni tribal governments.

Publications: A newsletter, *NACE News*.

Natural Resources Defense Council
40 West 20th Street
New York, NY 10011
(212) 727-2700

The Natural Resources Defense Council (NRDC) is a nonprofit organization of more than 170,000 members and a staff of lawyers, scientists, and other environmental specialists. The council is dedicated to protecting the world's natural resources and ensuring a safe and healthy environment for all people. It works through legal action, scientific research, advocacy, and public education. One of the council's most effective approaches has been to sue private companies that pollute the environment. In recent years, the NRDC has begun to take a greater interest in issues of environmental inequity. In 1993, it hired Vernice Miller to direct the council's Environmental Justice Initiative and in 1994 it featured a series of articles on environmental justice in its quarterly journal, *The Amicus Journal*.

Publications: A quarterly journal, *The Amicus Journa,* and a five-times-a-year newsletter, *NRDC Newsline.* In addition, the council publishes many books, pamphlets, reports, and other publications on topics such as *Out of Breath: Children's Health and Air Pollution in Southern California, A Guide to New York City's Reservoirs and Their Watersheds, After **Silent Spring**, One Year after Rio,* and *Harvest of Hope: The Potential for Alternative Agriculture to Reduce Pesticide Use.*

The Panos Institute
1717 Massachusetts Avenue, NW, Suite 301
Washington, DC 20036
(202) 483-0044
(202) 965-5198 (FAX)
e-mail: Panos@cais.com

The Panos Institute was created in 1986 to promote development that is sustainable socially, environmentally, and economically. It maintains three major centers, in Washington, D.C., Paris, and London. The two primary objectives of the organization are to strengthen the capacity for local communities to gather and report reliable information and to deepen public understanding of world issues crucial to humane sustainable development.

The organization is supported by bilateral and multilateral donors, nongovernmental charities, and private foundations, including the United States Agency for International Development; the U.S. Institute of Peace; the Swedish, Norwegian, and Danish development assistance agencies; and the Ford Foundation.

In the United States, the work of Panos is organized under two divisions, Regional Information Partnerships and Social Imperatives. Under the first category, the institute offers media training workshops, co-production of information materials, sponsorship of fellowships, and outreach and dissemination activities. Under the second category, it focuses on "misunderstood or under-reported issues," such as HIV/AIDS, population and consumption, and conflict and development.

Publications: Books such as *We Speak for Ourselves: Social Justice, Race, and Environment; We Speak for Ourselves: Population and Development, Miracle or Menace?; Biotechnology and the Third World;* and *Why People Grow Drugs;* teaching modules, such as *Environment and Development, Population and Development,* and *Conflict and Development;* and two periodicals, *EcoReports* and *SIDAmérica.*

People for Community Recovery*
13116 S. Ellis Ave.
Chicago, IL 60627
(312) 468-1645
(312) 468-8105 (FAX)

Established in 1982 for the purpose of educating the public through workshops and seminars about hazardous conditions in communities of color. Issues of concern: toxics, air pollution, waste disposal, wildlife, pesticides, lead, asbestos, recycling, parks and recreation, worker safety, housing, environmental justice, and community organizing.

Pesticide Education Center
942 Market Street, Suite 409
P. O. Box 420870
San Francisco, CA 94142-0870
(415) 391-8511
(415) 391-9159 (FAX)
e-mail: pec@igc.apc.org

Dr. Marion Moses, president and founder of the Pesticide Education Center (PEC), first worked with Cesar Chavez and the United Farm Workers as a full-time volunteer from 1966 to 1971. She later returned to work as Medical Director of the National Farm Workers Health Group from 1983 to 1986 before founding PEC in 1988. The purpose of the center is to educate the public about the hazards and health effects of pesticides and about the availability of nontoxic pest control methods that are safe for workers, consumers, and the environment.

The first project of the center was the production of videos and training manuals in English and Spanish to teach farmworkers about pesticides and how to protect themselves from pesticide exposure. Much of the center's work is carried out through other forms of public education, such as interviews in professional journals and the popular press.

Publications: A descriptive pamphlet, "Pesticide Education Center"; videos and training manuals mentioned above (also see chapter 7); reprints of articles about pesticide issues.

Santa Fe Health Education Project
P. O. Box 577
Santa Fe, NM 87504-0577
(505) 982-9520
(505) 982-3236 (FAX)

The Santa Fe Health Education Project is a small self-help and advocacy organization designed to provide health services in northern New Mexico. In addition, their large collection of health pamphlets has been sent to every state in the United States and to many users in Latin America and Europe. The project's efforts are designed to help individuals and self-help groups solve health problems and gain access to good medical care.

Publications: Five volumes of *Health Letters*, each volume containing 12 letters on topics such as pregnancy, immunizations, nutrition, animal bites, sexually transmitted diseases, amniocentesis, hormone replacement therapy, menstruation, menopause, mammograms, and microwave ovens. Each *Health Letter* is available for sale at 50¢ each, or 5 for $2.00

Sierra Club Legal Defense Fund
180 Montgomery Street, Suite 1400
San Francisco, CA 94104
(415) 627-6700
(415) 627-6740 (FAX)
e-mail: scldfsf@igc.apc.org

The Sierra Club Legal Defense Fund was created in 1971 to provide legal representation for public interest groups and conservation-minded individuals primarily and less commonly for trade associations and government agencies in legal actions and administrative appeals on a variety of issues, including air and water pollution, natural parks and forests, wilderness areas and seashores, toxic wastes, and environmental justice. The fund employs 36 attorneys in nine offices: San Francisco, Washington, D.C., Denver, Juneau, Seattle, Honolulu, Tallahassee, New Orleans, and Bozeman. In addition to its staff attorneys, the fund uses private attorneys who work on a *pro bono* or reduced-fee basis. In its 1994 Annual Report, the fund claims that "Increasingly we are considered the law firm for the environmental movement."

Examples of the work being done by the fund in the area of environmental justice include cases involving development of the predominantly minority Kingman Park section of southeast Washington, D.C., for an amusement park, freeway, and new football stadium; investigations of toxic waste regulations in the state of Mississippi for possible violations of the Civil Rights Act of 1964; action to ban the pesticide responsible for the largest number of farmworkers' hospitalizations; appeal of denial of political asylum to a Bulgarian dissident forced to work in a toxic industrial facility; and construction of a new uranium enrichment plant in the predominantly African American community of Homer, Louisiana.

Publications: A quarterly newsletter on environmental law, *In Brief; Docket: 1995*, an update containing examples of the kinds of cases being undertaken by the fund; and annual reports.

Southwest Network for Environmental and Economic Justice
117 7th Street, NW
P. O. Box 7399
Albuquerque, NM 87194
(505) 242-0416
(505) 242-5609 (FAX)
e-mail: sneej@ig.org

The Southwest Network for Environmental and Economic Justice was organized in April 1990 at a meeting of 30 southwestern organizations arranged by the SouthWest Organizing Project. The meeting included more than 80 Hispanic American, African American, and Native American activists from Oklahoma, Texas, New Mexico, Colorado, Arizona, Utah, Nevada, and California representing over 30 different organizations. The governing body of the network is a coordinating council, which was first elected at the 1990 meeting and which now serves as the network's board of directors. In 1992, a representative from the Tijuana area of Mexico was added to the board, making it the first organization in the United States of its kind to include a representative from Mexico on its board of directors.

In its mission statement, the network states that "As people of color, the Southwest Network recognizes that the demand for a safe, clean environment and workplace can only be achieved by building a multiracial and international movement that promotes environmental and economic justice." The work of the network is organized into major regional campaigns and projects: Environmental Protection Agency: Campaign for Accountability and Environmental Justice; Sustainable Communities; Border Justice; Sovereignty and Dumping on Native Lands; Youth Leadership and Development; and La Campaña Campesina: Farmworker Justice Against Pesticides.

Publications: Informational pamphlets, *Building a Net That Works*, *Southwest Network for Environmental and Economic Justice* (in

English and Spanish), and *Southwest Network for Environmental and Economic Justice: Celebrating the Spirit of Life—Weaving a Net that Works*.

SouthWest Organizing Project
211 10th St., SW
Albuquerque, NM 87102
(505) 246-8832
(505) 247-9972 (FAX)
e-mail: swop@igc.apc.org

The SouthWest Organizing Project was founded in 1981 as a statewide, multiracial, multi-issue, grassroots membership organization. One of its guiding principles is that people will live peacefully with each other only with the elimination of racism, sexism, and age and social class discrimination. The project is working for the self-determination of all peoples and for social, environmental, and economic justice at home and abroad. Among the activities in which the project has been engaged are the registration of voters, primarily in Hispanic American and African American communities; efforts to promote community control of housing, zoning, and local economic development; the promotion of health and safety concerns of workers; the development of educational materials; and opposition to the North American Free Trade Agreement. The project's position on environmental justice is outlined in its "Community Bill of Rights," which deals with issues of empowerment, information, and the responsibilities of industry and government.

Publications: A newsletter, *Voces Unidas*; a curriculum project, "Chicano History Teaching Kit" and a companion two-part video, "¡Viva La Causa! 500 Years of Chicano History; *Curriculum Guide for Elementary and Secondary School Teachers*, designed to accompany the book and video; a pictorial history, *500 Years of Chicano History in Pictures*; "Report on Interfaith Hearings on Toxic Poisoning in Communities of Color"; SWOP Bilingual Coloring Books; and a book, *Intel Inside New Mexico: A Case Study of Environmental and Economic Injustice*.

United Farm Workers of America, AFL-CIO
P. O. Box 62
Keene, CA 93531
(805) 822-5571, Ext. 3-142
(805) 822-6103 (FAX)

The United Farm Workers of America (UFWA) was established by Cesar E. Chavez in 1962. Chavez had been working since 1952 for the Community Service Organization, but had come to the conclusion that "to make basic changes in the lives of the poor, it was necessary to go to their work place and change conditions there." After a long summer of "house meetings" held in the homes of workers, Chavez called an organizational meeting of the UFWA in Fresno, California, on 30 September 1962.

Among the many activities through which the UFWA attempts to improve the lives of farmworkers are the union newspaper, *El Malcriado*; a farmworkers' credit union; a burial insurance program; the National Farm Workers Service Center; the Robert F. Kennedy Worker Medical Fund; the National Farm Workers Health Group; Agbayani Village, a farmworker retirement village; the Martin Luther King Jr. Farm Workers Fund; the Farm Workers Institute for Education and Leadership Development; the Education and Legal Defense Fund; the Marco Camacho Legal Corporation; and *La Unión del Pueblo Entero*, the latest benefits and services program offered by the UFWA.

In addition to the development of this broad range of services for farmworkers, UFWA has also been intensely active on the legislative front as well as instituting direct action, such as its notable grapes boycott, when such action seemed appropriate.

Publication: A newspaper, *El Malcriado*.

U.S. Environmental Protection Agency
Office of Environmental Justice (3103)
401 M Street S.W.
Washington, DC 20460
(202) 260-6357
(800) 962-6215
(202) 260-0852 (FAX)

The U.S. Environmental Protection Agency (EPA) established an Office of Environmental Justice in 1992 to coordinate the agency's efforts to address environmental justice issues. One of the first steps taken was to appoint an environmental justice coordinator at each of the EPA regional offices and its headquarters office. The agency also created the National Environmental Justice Advisory Council consisting of 25 members and six subcommittees, each with an additional 5–10 members appointed from key environmental justice constituencies. The council meets two to three times a year throughout the country in order to hear comments from local citizens and community groups on local issues related to environmental justice.

The EPA has also sponsored summer intern programs for the purposes of encouraging students to pursue a career in the environmental sciences. Two grant programs are also available to provide assistance of not more than $20,000 (in 1995) and to support partnerships between local groups and university personnel to work on environmental justice issues.

The agency provides an environmental justice hotline (1-800-962-6215) that provides information about environmental justice; offers technical support to communities attempting to reduce sources of risk to their people; supplies information about state and federal legislation dealing with environmental justice; and makes referrals to other government agencies and community groups that will provide further assistance with problems involving environmental justice.

Publications: An annual report that provides a detailed description of EPA activities in the field of environmental justice; a booklet "Environmental Justice Strategy: Executive Order 12898"; more than 40 other fact sheets, reports, bibliographies, interviews, brochures, pamphlets, and other publications.

The Video Project
5332 College Ave., Suite 101
Oakland, CA 94618
(510) 655-9050
(800) 4-PLANET (475-2638)
(510) 655-9115 (FAX)

The Video Project was created in 1983 as a nonprofit source for high-quality, affordable multimedia programs on issues critical to the fate of the planet and its people. Its founders were Vivienne Verdon-Roe, who won an Oscar for her film "Women—for America, for the World," and Ian Thiermann, who won an Oscar nomination for his film "In the Nuclear Shadow." The project distributes not only its own films, but also those of organizations such as the National Wildlife Federation, Marine Mammal Fund, League of Women Voters, Union of Concerned Scientists, and Zero Population Growth. Each year the project distributes over 10,000 programs to schools, colleges, community groups, public libraries, churches, businesses, government agencies, and individuals.

A special focus in the last few years has been the Project's Teaching the Next Generation Campaign, an effort through which 6,000 videos were donated to 2,000 schools, primarily in disadvantaged and culturally diverse communities. The majority of the films and videos made available by the project fall into the general categories of the environment, nuclear issues, war and peace, and human rights.

See chapter 7 for a list of the project's products in the field of environmental justice.

Publications: A catalog of films and videos "for a safe and sustainable world."

Washington Office on Environmental Justice
1511 K Street, N.W., Suite 1026
Washington, DC 20005
(202) 637-2467
(202) 637-9435 (FAX)

The Washington Office on Environmental Justice was created in 1994 to promote environmental and economic justice on behalf of multicultural, grassroots organizations and communities. The four major goals of the organization include the building and strengthening of regional networks, community groups, and organizations and their integration into mainstream decisionmaking on public policy matters; the improvement of communication and networking among environmental justice, grassroots, professional, scientific, and public policy organizations; the promotion of information discrimination, public awareness, and education; and the conduct of policy forums, training, and briefings on environmental justice issues. The areas of interest in which the office has chosen to work include transportation and energy; pollution prevention and environmental remediation; indigenous land rights and sovereignty; clean air and water; worker health and safety; job creation and technical assistance; military toxics and nuclear waste; and environmental education.

Publications: A newsletter, *The Washington Office on Environmental Justice Newsletter*.

West Harlem Environmental ACTion (WHE ACT)
271 West 125th Street, Suite 211
New York, NY 10027
(212) 961-1000
(212) 961-1015 (FAX)

The North River Sewage Treatment Plant in the West Harlem region of New York City was opened in April 1986 and, from the first day of operation, made itself known in the neighborhood because of its daily release of foul odors. Many residents of the area believed that the respiratory problems they experienced could be traced directly to the plant's emissions. Two years later, a small core of West Harlem residents had had enough and staged a protest against the plant, insisting that the city take action to bring its emissions under control. Three months after the initial protests, WHE ACT was formed with three key objectives in mind: to force the city of New York to correct problems at the sewage treatment plant; to obtain the right to take part in future planning and siting decisions for the neighborhood; and to bring issues of environmental justice before decisionmakers in the municipal government. They were eventually successful in reaching all three goals.

Today, WHE ACT is a community-based, nonprofit organization working to inform, educate, train, and mobilize the predominantly African American and Latino residents of northern Manhattan. The organization deals with any issue that affects the quality of life in the area: air and water pollution, housing problems, toxins, land use and open space, waterfront development and usage, sanitation, transportation, historic preservation, regulatory enforcement, and citizen participation in public policy-making.

Examples of the activities in which WHE ACT has been engaged are health surveys, a youth program to build awareness of environmental issues, a lead poisoning prevention program, a neighborhood environmental center, community forums, and a community environmental resources library.

Publications: A bilingual, bimonthly newspaper; numerous reprints of articles from local newspapers.

Working Group on Community Right-to-Know
218 D Street, S. E.
Washington, DC 20003
(202) 546-9707
(202) 546-2461 (FAX)

The Working Group on Community Right-to-Know is an affiliation of local, state, and national environmental groups concerned with the public's right to know about chemical hazards and toxic pollution. The organization, founded in 1987, provides technical assistance, carries out research, prepares reports, and offers referrals on issues such as toxic chemicals, right-to-know programs, and chemical accident prevention. Some of the groups affiliated with the working group are Clean Water Action, Environmental Defense Fund, Friends of the Earth, Greenpeace, Mineral Policy Center, National Coalition against the Misuse of Pesticides, Physicians for Social Responsibility, Sierra Club, and World Wildlife Fund.

Publications: A bimonthly newsletter, *Working Notes,* and 10 resource packets on issues such as "What Is the Emergency Planning and Right-To-Know Act?," "Chemical Accidents and Communities," "Plume Mapping," "Citizens Suits under EPCRA," "Toxics Emissions," and "Toxics-Use Reduction."

Selected Print Resources 6

Environmental justice is still a relatively young field. Insufficient time has passed to permit the development of an extensive literature base about the subject, as is now the case in other fields of social injustice or environmentalism. The bibliography listed here attempts to include the great majority of books that have been written so far on the subject of environmental racism, environmental equity, and environmental justice. It also includes a number of other sources to which scholars in the field of environmental justice themselves refer. In many cases, these sources were written before environmental justice self-consciously became a field of social and political action and of academic research. Yet, they often dealt with precisely the issues—disproportionate distribution of environmental impacts—with which the study and practice of environmental justice is concerned. On the other hand, many resources on important related topics—institutional racism in the United States, for example—have been omitted from the bibliography because they fail to deal specifically with the issue of environmental racism. Books providing more extensive bibliographies on such ancillary topics are indicated in the following bibliography.

Alston, Dana, ed. *We Speak for Ourselves: Social Justice, Race and the Environment*. Washington, DC: Panos Institute, 1990. 40 pp. ISBN 1-879358-01-8.

This paperback publication describes "the marriage of the movement for social justice with environmentalism,"according to the editor's introduction. Most of the 13 articles in the book describe specific instances in which activists have worked to deal with environmental issues in communities of color. Other articles describe the issues involved in bringing together people of color and people from low-income groups with more traditional (and, usually, mostly white) environmental groups. The final chapter describes the international aspects of the environmental justice issue. A two-page resource guide is included; unfortunately, a significant number of the organizations listed no longer exist.

Americans for Indian Opportunity. *Messing With Mother Nature Can Be Hazardous To Your Health*. Washington, DC: [no date]. 223 pp. [paginated by section].

This book is a report of the activities of Americans for Indian Opportunity designed to assess the environmental health impacts of development on Native American communities and the roles of various governmental agencies charged with the responsibilities for various aspects of environmental protection and individual safety. The overall project had five major objectives: increasing the awareness of Native American tribes about environmental health effects of development; developing information and documentation needed to assist Native American communities in protecting their own health and environment; developing alternative methods for organizing Native American community environmental health protection systems; increasing awareness of governmental agencies about environmental health issues in Native American communities; and establishing communication among Native American and non-Native American individuals and communities about environmental health issues. The book summarizes activities that have been developed to accomplish each of these objectives.

Angel, Bradley. *Toxic Threat to Indian Lands: A Greenpeace Report*. San Francisco: Greenpeace, 1991. Unpaginated [19 pp.].

This short report argues that the waste disposal industry has targeted lands owned by Indian Nations (Tribes) as promising areas in which to dump waste materials, some of them toxic, radioactive, or otherwise hazardous. The main body of the report lists all instances known to the author in which there have been specific examples to support this thesis.

Belliveau, Michael, Michael Kent, and Grant Rosenblum. *Richmond at Risk: Community Demographics and Toxic Hazards from Industrial Polluters*. San Francisco: Citizens for a Better Environment, February 1989. 130 pp.

This study was completed by Citizens for a Better Environment (CBE), a public interest group that uses research, policy advocacy, litigation, organizing, and education to prevent and reduce toxic hazards to human health and the environment arising from pollution of air, water, and land resources in major urban areas of California. The study examined a number of issues confronting the Richmond area of Contra Costa County in the East Bay region of the San Francisco Bay community. Included among these issues were routine toxic pollution, toxic chemical accidents, environmental racism, toxic hazard assessment, promises of pollution prevention, and state and national implications of environmental hazards in the Richmond area.

The study resulted in a number of important conclusions. In the area of routine pollution, for example, researchers concluded that "Significant amounts of poorly regulated toxic pollutants are routinely generated by industrial facilities in the Richmond area in the form of air emissions, waste water discharges, and hazardous waste." With regard to the issue of environmental racism, the report indicated that "Minority residents of the Richmond area bear a disproportionate share of toxic chemicals because of the high concentration of industrial facilities located in close proximity to predominantly lower income, Black and Hispanic neighborhoods." In order to deal with this problem, the CBE made recommendations, such as installing more buffer zones between industrial sources of pollution and residential neighborhoods, requiring stringent land use permits for existing industrial facilities, and initiating cooperative efforts between public and private sector institutions to ensure that all new industrial facilities are either nonpolluting or designed to reduce hazard waste emissions to as low a level as possible.

Berry, Brian J. L., et al. *Land Use, Urban Form and Environmental Quality*. Chicago: Department of Geography Research Paper No. 155, University of Chicago, 1974. 441 pp.

This study was carried out for the Office of Research and Development of the U.S. Environmental Protection Agency in an effort to evaluate the ways various types of land use produce a variety of environmental effects. The primary environmental issues investigated were air and water quality, solid waste disposal, noise pollution, pesticide use, and radiation. The study is of interest because of the attention it gives differential environmental impacts among various categories of people, such as different racial and ethnic groups and populations of different socioeconomic status.

Berry, Brian J. L., ed. *The Social Burdens of Environmental Pollution: A Comparative Metropolitan Data Source*. Cambridge, MA: Ballinger Publishing Company, 1977. 613 pp. ISBN 0-88410-427-3.

This study was undertaken for the U.S. Environmental Protection Agency in an attempt to find patterns of disproportionate environmental impact on people of various races and income. The researchers studied surface water quality; air quality; urban noise; and solid waste generation, collection, and disposal in 12 metropolitan regions (Baltimore, Birmingham, Cincinnati, Denver, Jacksonville, Oklahoma City, Providence, Rochester, San Diego, Seattle, St. Louis, and Washington, D.C.). In addition, they conducted a more detailed study of water quality in the metropolitan Chicago area. This study contains a large amount of data on these topics, presented in well-drawn maps and clear tables. The authors conclude from their study, among other things, that "the minority poor, living in the oldest highest-density inner-city neighborhoods, are afflicted with the greatest pollution burdens—joined in some cities by the more-elderly apartment-living affluent. But these same inner city neighborhoods also are beset by a complex of other ills related to poverty and poor housing, including greater risk of rat bites, and of poisonings by the rodenticides and pesticides used to keep unwanted pests in check."

Bryant, Bunyan, ed. *Environmental Justice: Issues, Policies, and Solutions*. Washington, DC: Island Press, 1995. 278 pp. ISBN 1-55963-416-2.

This collection of articles covers a broad range of topics, including "Health-Based Standards: What Role in Environmental Justice?" "Environmental Justice Centers: A Response to Inequity," "Environmentalists and Environmental Justice Policy," "Residential Segregation and Urban Quality of Life," "The Net Impact of

Environmental Protection on Jobs and the Economy," "Towards a New Industrial Policy," "Environmental Equity and Economic Policy: Expanding the Agenda of Reform," "Indigenous Nations: Summary of Sovereignty and its Implications for Environmental Protection," "Sustainable Agriculture Embedded in a Global Sustainable Future: Agriculture in the United States and Cuba," and "Rethinking International Environmental Policy in the Late Twentieth Century."

Bryant, Bunyan, and Paul Mohai. *Race and the Incidence of Environmental Hazards: A Time for Discourse*. Boulder, CO: Westview Press, 1992. 251 pp. ISBN 0-8133-8513-X.

This collection of 14 articles by a variety of writers is divided between discussions of specific examples of environmental racism ("Environmentalism and Civil Rights in Sumter County, Alabama," "Invitation to Poison? Detroit Minorities and Toxic Fish Consumption from the Detroit River," and "Uranium Production and Its Effects on Navajo Communities along the Rio Puerco in Western New Mexico") and more general discussions of fundamental issues in environmental justice ("Toxic Waste and Race in the United States," "Can the Environmental Movement Attract and Maintain the Support of Minorities?" and "Environmental Blackmail in Minority Communities"). Of particular value is the extensive list of references provided for each of the individual chapters—more than 30 pages of books, articles, and reports overall.

Bullard, Robert D., ed. *Confronting Environmental Racism: Voices from the Grassroots*. Boston: South End Press, 1993, 259 pp. ISBN 0-89608-447-7.

Some of the 12 chapters in this anthology describe specific instances of environmental racism, including discussions as to how individuals and organizations dealt with those instances. Examples of these cases are chapters on "Race and Waste in Two Virginia Communities," "Environmental Politics in Alabama's Blackbelt," "Sustainable Development at Gannados del Valle," and "Nature and Chicanos in Southern Colorado." Other chapters consider broader issues in environmental justice, such as Bullard's introductory chapter on "Anatomy of Environmental Racism and the Environmental Justice Movement," and later chapters on "Environmentalism and the Politics of Inclusion," "Coping with Industrial Exploitation,"and "Farmworkers and

Pesticides." The bibliography provides an extraordinarily complete and useful list of books, articles, and reports dealing with all aspects of environmental justice, racism, environmental issues, and related topics.

Bullard, Robert D. *Dumping in Dixie: Race, Class, and Environmental Quality*. Boulder, CO: Westview Press, 1994. 195 pp. ISBN 0-8133-1962-5.

This book is one of the seminal works in the environmental justice movement. It arose out of Bullard's research in Houston on the spatial distribution of municipal solid waste disposal sites. Bullard then extended his research to other areas in which environmental racism appeared to exist, specifically, West Dallas, Texas; Institute, West Virginia; Alsen, Louisiana; and Emelle-Sumter County, Alabama. This book reports the results of his research in these areas in addition to providing a general analysis of the nature of environmental racism and the ways in which it can be combated. The author explains in the preface to the book that his goal is to "write a readable book that might reach a general audience while at the same time covering uncharted areas of interest to environmentalists, civil rights advocates, political leaders, and policy makers."

Bullard, Robert D. *People of Color Environmental Groups, 1994-95 Directory*. Atlanta: Environmental Justice Resource Center, Clark Atlanta University, 1994 (updated from 1992 edition). 194 pp.

This directory lists more than 300 people of color groups from 40 states, the District of Columbia, Puerto Rico, Canada, and Mexico. The listing for each organization includes its address and telephone number, contact person, missions, issues with which the organization is concerned, activities, and statistical and background information, such as number of paid staff, members, volunteers, constituency served, geographic focus, and year founded. The directory is important because it brings together in a single volume the most complete set of data about organizations at all levels—local, regional, and national—dealing with environmental justice.

Bullard, Robert D., ed. *Unequal Protection: Environmental Justice and Communities of Color*. San Francisco: Sierra Club Books, 1994. 392 pp. ISBN 0-87156-450-5.

In his introduction to this book, the editor points out that the major theme of the volume is "the cultural diversity of the environmental justice movement." He reminds readers that its contributors have come from an array of fields, including academia, journalism, and law. Part I of the book deals with a historical background of the environmental justice movement, reviewing two early cases that helped define the movement—at Indian Creek, Alabama, and Warren County, North Carolina. Part II describes cases that have arisen in "sacrifice zones," areas where high concentrations of industrial pollution are to be found. Among the examples discussed in this section of the book include a Superfund site in Texarkana, a lead smelter in Dallas, and the area of the Mississippi River valley known as "Cancer Alley." Part III of the book focuses on the alliances that have developed between mainstream environmentalists and grassroots organizations interested specifically in issues of environmental justice. Among the organizations discussed are the Southwest Organizing Project, Concerned Citizens of South Central Los Angeles, and Mothers of East Los Angeles.

Center for Investigative Reporting and Bill Moyers. *Global Dumping Ground*. Cambridge: The Lutterworth Press, 1991. 152 pp. ISBN 0-932020-95-X.

This book was written in conjunction with a PBS documentary special on the subject of the international trade in hazardous waste disposal. The authors point out that developed nations export millions of tons annually of used car batteries, dry cleaning fluids, pesticides, and other hazardous wastes to some of the poorest nations in the world. In addition to providing a number of case studies on this topic, the book outlines some ways in which the problem of hazardous waste dumping can be remedied. The book provides a valuable list of books, documents, articles, and films dealing with the topic, as well as a list of organizations concerned with the issue.

Churchill, Ward. *Struggle for the Land: Indigenous Resistance to Genocide, Ecocide and Expropriation in Contemporary North America*. Monroe, ME: Common Courage Press, 1993. 472 pp. ISBN 1-56751-000-0 (paperback).

In his introduction, the author explains that his purpose in writing this book is to help in "reforging popular consciousness of things Indian in modern North America, and thereby to facilitate the emergence of genuine alliances between Indians and non-Indians." The book is divided into four major parts: "American 'Justice'," a general historical background of the relationship between Native Americans and white people who arrived after 1492; "In Struggle for the Land," specific examples of cases in which Indians have been expelled from their native lands; "Other Battles," general articles on issues between Indians and other North Americans; and "I Am Indigenist," a statement of the author's philosophy about his own role in the kinds of battles described in other sections of the book.

Commission for Racial Justice. United Church of Christ. *Toxic Wastes and Race in the United States: A National Report on the Racial and Socioeconomic Characteristics of Communities with Hazardous Wastes Sites*. New York: Public Data Access, Inc., 1987. 69 pp.

This report is an important document produced in the early history of the environmental justice movement. In January 1986, two studies were initiated by the Commission for Racial Justice of the United Church of Christ to determine the extent to which various minority groups are exposed to hazardous wastes in their communities. One study dealt with commercial hazardous waste facilities, and the other on uncontrolled toxic waste sites. The major conclusions and recommendations resulting from this study are reprinted in chapter 4 of this book.

Costner, Pat, and Joe Thornton. *Playing with Fire*. Washington, DC: Greenpeace, 1990. 64 pp.

For many years, American communities have been finding it increasingly difficult to locate sites on which to build waste disposal facilities. One response to this problem has been a move toward the incineration of wastes. A number of authorities have argued that burning up wastes has many advantages over burying them underground in landfills. This report was prepared in response to that growing movement in support of waste incineration.

The authors devote the first four chapters of their report to an analysis of the science and technology of waste disposal by incineration. They discuss techniques of incineration, products of complete and incomplete combustion of wastes, and the special problem of incineration of metal-containing wastes. Chapter five, then, discusses the health and environmental impacts of incinerator releases. Finally, chapter six provides a review of the social, political, and economic issues involved in the use and siting of incineration plants. The final section, "Communities near Hazardous Waste Incinerators—Hosts or Hostages?," is especially relevant to the discussion of environmental equity issues.

The report is enhanced by four appendices tabulating the characteristics and locations of existing commercial hazardous waste incinerators and cement/aggregate kilns currently burning wastes and 22 tables of useful and interesting data about hazardous waste incineration.

Davies, J. Clarence III, and Barbara S. Davies. *The Politics of Pollution*, 2nd edition. Indianapolis: Pegasus, 1975. 254 pp. ISBN 0-672-63720-0 (paperback).

The authors discuss how public policy on pollution is made and how various interest groups attempt to influence that process. The book is divided into three parts, the first of which describes the nature of the pollution problem and the growing interest in that issue expressed in the 1970s. Part II analyzes the various forces—the executive, legislative, and judicial branches of the federal government; public opinion; special interest groups; and state and local governments—that have had an impact on the making of pollution policy. The final part of the book examines the processes by which policy on pollution issues is actually made.

Edelstein, Michael R. *Contaminated Communities: The Social and Psychological Impacts of Residential Toxic Exposure*. Boulder, CO: Westview Press, 1987. 217 pp. ISBN 0-8133-7447-2.

The author is an environmental psychologist who has spent years studying the social and psychological effects on ordinary citizens who have been exposed to toxic and hazardous wastes. His thesis is that the social sciences can offer guidance in helping people whose lives have been disrupted by environmental disasters, such as the Love Canal incident. After providing a general introduction to the subject and reviewing a typical case study of a contaminated community (Jackson, New Jersey), the author discusses the cognitive, psychological, and social adjustments that people have to make to toxic exposures, the responses that governments provide to citizen complaints, and the mechanisms by which citizens can respond to industry and to government. An exhaustive bibliography is provided.

Finkel, Adam M., and Dominic Golding, eds. *Worst Things First? The Debate over Risk-Based National Environmental Priorities.* Washington, DC: Resources for the Future, 1994. 348 pp. ISBN 0-915707-74-8.

From November 15 to 17, 1992, a conference of more than 100 individuals was held in Annapolis, Maryland, to discuss methods by which the national environmental agenda should be established. The conference was sponsored by The Center for Risk Management at Resources for the Future. The motivation for the conference was the decision by the U.S. Environmental Protection Agency to begin setting national priorities on a "rational" basis, using risk assessment and expert judgment, rather than solely on the basis of political or crisis-oriented criteria.

The book contains 17 papers presented at the conference, as well as the keynote address by Alice Rivlin, Director of the Office of Management and Budget, and concluding summaries and remarks by the editors.

Part Three of the book contains three paradigms for approaches to setting national environmental priorities; the prevention paradigm, the industrial transformation paradigm, and the environmental justice paradigm. The last of these contains papers by Robert D. Bullard, "Unequal Environmental Protection: Incorporating Environmental Justice in Decision Making," and Albert L. Nichols, "Risk-Based Priorities and Environmental Justice."

Florini, Karen, George D. Krumbhaar, Jr., and Ellen K. Silbergeld. *Legacy of Lead: America's Continuing Epidemic of Childhood Lead Poisoning.* Washington, DC: Environmental Defense Fund, March 1990. 46 pp., appendices.

This report is a succinct and thorough review of the problem of lead in the environment and its health effect on humans, particularly young children. The authors review the evidence about toxicity, the ways by which it reaches and enters the body, the number of individuals affected by lead poisoning, treatment and prevention options, and governmental action that has and should be taken. The final two chapters of the report outline a proposal for dealing with the problem of lead in the environment. Of particular interest is a lengthy appendix tabulating the number of children thought to be at risk for lead exposure in small, medium, and large Standard Metropolitan Statistical Areas.

Fumento, Michael. *Science under Siege: Balancing Technology and the Environment*. New York: Morrow, 1993. 448 pp. ISBN 0-688-10795-8.

The author argues that many concerns expressed about environmental issues are exaggerated by environmentalists, politicians, and members of the media who ignore scientific facts and use scare tactics to promote environmental policies. He says that this book is "a plea for rational public policy" based on sound and dependable scientific information, and not on media campaigns. The book focuses on a number of specific environmental issues, including the pesticide Alar, dioxin, Agent Orange, food irradiation, electric and magnetic field effects, video display terminals (VDTs), and gasohol. Chapter 12, "A Closer Look at the Besiegers," deals with those whom Fumento believes are responsible for the "state of siege"under which some Americans now feel that they live. While the book does not deal specifically with environmental justice issues, it does speak to the general question of how environmental risks should be evaluated by society as a whole.

Gaventa, John, Barbara Ellen Smith, and Alex Willingham, eds. *Communities in Economic Crisis: Appalachia and the South*. Philadelphia: Temple University Press, 1990. 301 pp. ISBN 0-87722-650-4.

Two chapters in this book deal specifically with environmental justice issues. "Environmentalism, Economic Blackmail and Civil Rights" discusses the dilemma in which many minority individuals find themselves confronted with the opportunity of working in a hazardous workplace or having no job at all. "Economic Slavery or Hazardous Wastes: Robeson County's Economic Menu" discusses the long history of hazardous industries that have located in Robeson County, with special attention to the battle to keep the GSX Corporation from locating a new facility on the Lumbee River.

Goldman, Benjamin A. *Not Just Prosperity: Achieving Sustainability with Environmental Justice*. Washington, DC: National Wildlife Federation, Corporate Consumer Council, 1993. 49 pp.

This report analyzes the role of environmental justice in the question of sustainable development. It was prepared for the National Wildlife Federation's Corporate Conservation Council's conference on "Synergy '94: The Community Responsibilities of Sustainable

Development," held in February 1994. Part I of the report provides an exhaustive review of research that had been conducted on the issue of environmental racism and environmental equity to date. Part II of the report then analyzes the implications of this research for sustainable development. The report is an excellent overall summary of the objective evidence that has been collected so far on the extent and nature of environmental inequities. An extensive and valuable bibliography is also included in the report.

Goldman, Benjamin, and Laura Fitton. *Toxic Wastes and Race Revisited*. Washington, DC: The Center for Policy Alternatives, 1994. 27 pp.

This report was sponsored by the United Church of Christ Commission for Racial Justice, the Center for Policy Alternatives, and the National Association for the Advancement of Colored People in an attempt to discover what changes had taken place with regard to toxic waste landfills since the historic 1987 study by the Commission for Racial Justice. The authors note in their introduction to this report that the widespread existence of commercial toxic waste sites (530 examined for this study) at least partially represents a more aggressive effort on the part of the federal government to identify toxic chemicals for which disposal sites must be located. That effort, in turn, has been mandated by the 1976 Resource Conservation and Recovery Act.

In general, the study found that the disproportionate exposure of people of color to hazardous waste sites had actually increased in the years between 1980 and 1993 at national, state, and regional levels. The concentration of people of color living in ZIP code regions with commercial waste facilities, for example, had increased from 25 percent to 31 percent during that time period. The study also found that the disproportionate siting of hazardous waste facilities appeared to correlate less well with socioeconomic measures than with racial measures.

Goldman, Benjamin. *The Truth about Where You Live: An Atlas for Action on Toxins and Mortality*. New York: Random House, 1991. 416 pp. ISBN 0-8129-1898-3 (paperback).

Under the Freedom of Information Act, the author and his colleagues obtained voluminous data about a variety of environmental hazards to which American citizens are exposed and assessed the degree of exposure to such hazards by residents in every county and parish in the United States. The result is a book consisting of hundreds of maps and tables describing the extent to which various regions of the nation are at risk for health problems such as heart disease, birth defects, all forms of cancers, infectious diseases, industrial toxins, and nonindustrial pollution. Chapter 6, "Environmental Justice," focuses on the disproportionate extent to which minorities and poor communities are exposed to such hazards. The book is a critical reference for those interested in studying the factual data on which claims of environmental inequities are often based.

Goldsmith, Edward, Nicholas Hildyard, Patrick McCully, and Peter Bunyard. *The Imperiled Planet*, Cambridge, MA: MIT Press, 1990. 288 pp. ISBN 0-262-07132-0.

The hypothesis proposed in this book is that the planet (analyzed here by means of the "Gaia" hypothesis) is seriously threatened by technological developments of the last century. The authors suggest, for example, that "we are pushing natural processes beyond their capacity to maintain an atmosphere fit for higher forms of life." The major portion of the text is devoted to detailed analyses of the status of the atmosphere, forests, agricultural lands, rangelands, rivers, groundwater, wetlands and mangroves, coasts and estuaries, seas and oceans, coral reefs, islands, mountains, deserts, Antarctica, and the Arctic. The final section of the book includes four chapters on "The Diminishing Quality of Life," "The Future in Prospect," "The Dynamics of Destruction," and "Solutions for Survival." The book is lavishly illustrated with more than 200 color photographs.

Gould, Jay M., and Alice Tepper Marlin, ed. *Quality of Life in American Neighborhoods: Levels of Affluence, Toxic Waste, and Cancer Mortality in Residential ZIP Code Areas*. Boulder, CO: Westview Press, 1986. 402 pp. ISBN 0-8133-7187-2.

Most of this book consists of 35,000 ZIP code areas with 50 or more households with 13 pieces of statistical information about each area, including population, percentage of white residents, number of households, percentage of households owner occupied, mean income per household, mean monthly rent, mean home value, percentage of residents under the age of 5 and over the age of 65, percentage of white males, percentage of white males over the age of 65, number of toxic waste sites, and the amount of toxic waste relative to the U.S. average. Three "striking findings" are reported as a result of this study. One is the "astonishingly wide variation" in incomes in different ZIP code areas. A second is the "disproportionately high levels" of toxic waste concentrations in ZIP code areas with low mean income. A third is that regions of high toxic waste concentration may also border closely on ZIP code areas characterized by high incomes, high rentals, and high home values.

Greenberg, Michael R., and Richard F. Anderson. *Hazardous Waste Sites: The Credibility Gap.* New Brunswick, NJ: Rutgers University Center for Urban Policy Research, 1984. 276 pp. ISBN 0-88285-102-0.

"Mutual distrust is probably a charitable way of characterizing the relationship between those who are the sources of hazardous wastes and those who oversee their activities. Furthermore, it appears that a large segment of the public trusts neither party to protect them." These two sentences outline the premise of this book and the problem with which it attempts to deal. The main body of the text is devoted to dealing with six issues about hazardous wastes: (1) definitions and sources of hazardous wastes; (2) legal and technical controls available for dealing with hazardous wastes; (3) the effects of hazardous wastes on human health and the environment; (4) the "legal, economic, and political realities of searching for new sites (for hazardous waste disposal) in the United States"; (5) technological methods for improving the credibility of the siting process; and (6) public and private actions that can close the credibility gap about hazardous wastes. The book contains a number of useful tables and maps.

Greve, Michael S., and Fred L. Smith, Jr., eds. *Environmental Politics: Public Costs, Private Rewards.* New York: Praeger, 1992. 212 pp. ISBN 0-275-94238-4 (paperback).

The contributors to this volume take the common view that the way in which environmental issues are handled by the American political system often produces results exactly contrary to those originally intended. The titles of some of the chapters reflect this view: "Clean Fuels, Dirty Air"; "Pollution Deadlines and the Coalition for Failure"; and "Ozone Layers and Oligolopy Profits." In his conclusion to the book, Fred L. Smith Jr. writes that many of the actions of the Environmental Protection Agency have "conform[ed] to the traditional political pattern of favoring special interest under the guise of serving the public interest."

The book is a valuable resource for introducing readers to the complex economic and political context within which many environmental problems—including those that arise under the topic of environmental racism—are handled by the traditional American political process.

Grinde, Donald, Jr., and Bruce Johansen. *Ecocide of Native America: Environmental Destruction of Indian Lands and Peoples*. Santa Fe, NM: Clear Light, 1995. 310 pp. ISBN 0-940666-52-9.

The authors point out the irony of the fact that Native Americans, who may have one of the most sensitive ecological philosophies of any peoples on Earth, have become the worst victims of the United States' assault on the natural environment. The book outlines the historical beliefs that Native Americans have held about the Earth and about the place of humans in the natural environment. The authors then describe a number of cases in which Native American societies have been exposed to environmental insults from mining companies, waste disposal operations, and other abusers of the environment. These case studies are selected from a wide historical period, ranging from the pre-Columbia period to the modern day. The book is valuable in its attempt to present a revisionist view of the relationship among Native Americans, colonists from Western Europe and their descendants, and the environment.

Harrison, David. *Who Pays for Clean Air: The Cost and Benefit Distribution of Federal Automobile Emission Controls*. Cambridge, MA: Ballinger Publishing Company, 1975. 167 pp. ISBN 0-88410-451-6.

The study that led to this book was completed more than a decade before the term "environmental justice" had even been coined. It provides some of the earliest evidence that environmental costs (the costs of air pollution, in this case) are not shared equally by all economic classes, but are concentrated within certain income groups, namely those with lower income. As an example of the conclusions reached by the author, he points out that "the current auto emission control scheme results in undesirable impacts for households in the suburban areas of American SMSAs [Standard Metropolitan Statistical Areas]. Suburban residents in all population size groups on average obtain low benefits and pay high costs. Moreover, these costs are borne quite heavily by lower income groups."

Hofrichter, Richard, ed. *Toxic Struggles: The Theory and Practice of Environmental Justice*. Philadelphia: New Society Publishers, 1993. 260 pp. ISBN 0-8657-1270-0.

The essays that constitute this book are presented, according to the editor, from a "perspective that places environmental issues in a larger context of struggles for social change, locally and globally." The book was written as a project of the Center for Ecology and Social Justice in Washington, D.C. The book is divided into two major parts, the first providing a theoretical perspective on the context of environmental justice. Chapters deal with topics such as "Capitalism and the Crisis of Environmentalism," "Anatomy of Environmental Racism," "The Promise of Environmental Democracy," "Creating a Culture of Destruction: Gender, Militarism, and the Environment," and "Feminism and Ecology." Part Two deals with practical issues and includes chapters such as "A Society Based on Conquest Cannot Be Sustained: Native Peoples and the Environmental Crisis," "Blue-Collar Women and Toxic-Waste Protests: The Process of Politicization," "Ecofeminism and Grass-Roots Environmentalism in the United States," "Farm Workers at Risk," "Corporate Plundering of Third-World Resources," and "Trading Away the Environment: Free-Trade Agreements and Environmental Degradation."

Hoyt, Homer. *The Structure and Growth of Residential Neighborhoods in American Cities*. Washington, DC: Federal Housing Administration, 1939. 178 pp.

This study has some historical interest in that it was one of the first efforts by the federal government to obtain, analyze, and publish data on the way in which cities and neighborhoods develop. Although certainly not a priority of the study, the unequal distribution of benefits and burdens in a community clearly shows through in the maps, tables, and data presented in this report.

Kasperson, Roger E., ed. *Equity in Radioactive Waste Management*. Boston: Oelgeschlager, Gunn, and Hain, 1983. 381 pp. ISBN 0-89946-055-0.

The issue of radioactive waste management could well become "the most intractable" of all environmental issues, according to the editor of this volume. The problem is that well into the second-half century of the atomic age, there is still no national policy in the United States to deal with radioactive wastes produced as a result of weapons research and development and of nuclear power production. The purpose of this book is to examine how past and current radioactive waste disposal practices affect various parts of society.

After an introductory section, "The Problem in Perspective," the book is divided into four major parts dealing with "The Locus Problem," "The Legacy Problem," "The Labor Problem," and "Toward Public Policy." The first three of these sections deal not only with theoretical and policy issues, but also with specific case studies involving radioactive waste disposal. The final chapter suggests some proposals for dealing with this issue in a socially, politically, and economically equitable fashion.

Kazis, Richard and Richard Grossman. *Fear at Work: Job Blackmail, Labor and the Environment*. Philadelphia: The Pilgrim Press, 1982.

Although now a decade and a half old, this book remains one of the standards in the field of "job blackmail,"an issue that remains as current today as it was when the book was first written. The authors, staff members at Environmentalists for Full Employment, present examples of instances in which corporations have threatened to close down factories or mines because the environmental and regulatory costs of keeping them open was supposedly too great for them to pay. As a result, workers have been confronted with the dilemma of supporting facilities that they know to be dangerous to their health and to the environment and maintaining jobs, or taking stands against such facilities and losing their jobs. In such cases, workers and environmentalists have traditionally ended up on opposite sides of issues such as pollution control improvements in factories, expanded regulatory protections, and the siting of hazardous waste sites.

In fact, as the authors demonstrate, the arguments made by corporations are often specious, using their objections to environmental regulations as a cover for business decisions that they might have had to make anyway. More commonly, they have simply shifted the burden of operating an unsafe industrial facility from themselves to their workers.

The dichotomy between protections for the environment and for occupational safety on the one hand and economic efficiency on the other is, in any case, the authors claim, a false one. Indeed, a healthy workplace, a healthy environment, and a healthy economy are not only *not* mutually exclusive, but instead can develop together fruitfully.

LaBalme, Jenny. *A Road to Walk: A Struggle for Environmental Justice*. Durham, NC: Regulator Press, 1987. 28 pp.

This short book tells the story of an attempt to prevent the construction of a dump for polychlorinated biphenyl (PCB) wastes in Warren, North Carolina. The book is effectively illustrated with more than 30 photographs documenting the protest movement that developed.

Lake, Elizabeth E., William M. Hanneman, and Sharon M. Oster. *Who Pays for Clean Water?: The Distribution of Water Pollution Control Costs*. Boulder, CO: Westview Press, 1979. 244 pp. ISBN 0-89158-586-9.

This study attempts to discover how costs of water pollution imposed as a result of the 1972 Amendments to the Water Pollution Control Act, P.L. 92-500, will be distributed throughout the U.S. population. The report presented here is an exhaustive study of how those costs will be distributed as a result of changes in both public policy and public services and increased costs to industry. The authors conclude that, in general, the costs of water pollution are *not* regressive and that higher income groups will pay proportionately more than their share of new water pollution control costs. They also point out, however, that, as a group, African Americans will pay proportionately more than their share of these costs because they constitute such a large fraction of lower income groups. More than 100 charts and tables illustrate the findings reported in the study.

Makhijani, Arjun. *From Global Capitalism to Economic Justice: An Inquiry into the Elimination of Systemic Poverty, Violence and Environmental Destruction in the World Economy*. New York: Apex Press, 1992. 176 pp. ISBN 0-945257-41-4.

The author's purpose in this book is to analyze the current economic system found in the world, a system he claims is based on war, capitalism, and colonialism, and to lay out a different type of world system that will produce "justice, peace, and environmental harmony." The key to the system that he describes is economic democracy and economic justice.

Mann, Eric. *L.A.'s Lethal Air: New Strategies for Policy, Organizing, and Action*. Los Angeles: Labor/Community Strategy Center, 1991. 80 pp. ISBN 0-96298-130-3.

The author points out how corporate and government policies operate to locate polluting facilities in the Los Angeles area, accounting for its nickname *valley of smoke*. He places his story in a broader context of social, racial, and economic justice and outlines a strategy by which such issues can be analyzed and attacked.

Meyer, Art and Jocele Meyer. *Earth Keepers: Environmental Perspectives on Hunger, Poverty, and Injustice*. Scottdale, PA: Herald Press, 1991. 264 pp. ISBN 0-8361-3544-X.

The authors argue that "problems of world hunger, poverty, conflict, and injustice are intricately intertwined with environmental issues" and that churches need to become more actively involved in solving problems of ecology and hunger. This book consists of essays and discussions the authors have had on these topics over the seven years of their work with the Office of Global Education of the Mennonite Central Committee U.S. Peace Section.

The book is divided into five major sections: "Stewardship of Creation—Earthkeeping," "Environmental Degradation," "Environment and Conflict," "Food and Energy," and "Sustainable Agriculture."A section "For Further Study" is provided, but no index is available.

Norton, Bryan G. *Toward Unity among Environmentalists*. New York: Oxford University Press, 1991. 287 pp. ISBN 0-19-506112-8.

This book is a useful reference about the origins of mainstream environmentalism, the stages through which that movement has gone, the pressures that have arisen between "traditional" environmentalists and "Earth Day" environmentalists, and the ways in which accommodations can be made among the various forms of environmentalism.

The value of the book for those interested in environmental justice lies not only in the historical and contemporary analyses presented, but also in helping the reader better understand the kinds of issues confronting those interested in the environment, whatever their political perspective.

Norris, Ruth, ed. *Pills, Pesticides & Profits: The International Trade in Toxic Substances*. Croton-on-Hudson, NY: North River Press, 1982. 167 pp. ISBN 0-88427-050-5.

The three contributors to this book are staff members of the Natural Resources Defense Council and an independent film producer.

They write about international policies and practices regarding trade in substances such as pesticides, pharmaceuticals, consumer products, and hazardous wastes. Chapter Five deals with "The Search for Solutions." They suggest that "we are now witnessing only the early signs of a growing problem—one in which increasing production of chemicals in the more highly developed countries will inevitably lead to the search for markets in the less developed parts of the world." One of the problems with this vision of the view, they point out, is that multinational firms "appear to thrive in a climate of 'benign' neglect by the governments of developing countries." Based on events that have transpired in years since the book was written, the contributors appear to have been on target with this prediction.

In addition to the main body of the book, some useful, if somewhat outdated, appendices are included. The first of these contains the transcript of a 1981 PBS broadcast, "Pesticides and Pills: FOR EXPORT ONLY." The second contains a number of useful and interesting tables on topics such as the 50 largest transnational pharmaceutical companies, foreign sales trends for 14 major U.S. pharmaceutical companies, and double standard cases in which the distribution, use, and sale of products are banned in one country and permitted in another.

Petrikin, Jonathan. *Environmental Justice*. San Diego: Greenhaven Press, 1995. 128 pp. ISBN 1-56510-264-9 (paperback).

This book is part of Greenhaven's "At Issue"series. It contains nine articles that argue various positions in the environmental justice issue. For example, an article by Paul Mohai and Bunyan Bryant claims that "Demographic Studies Reveal a Pattern of Environmental Injustice," while a following article by Douglas L. Anderton and his colleagues claims that "Studies Used to Prove Charges of Environmental Racism Are Flawed." The book is valuable not only because it presents opposing views on environmental justice within a single text, but it also contains a useful bibliography and list of organizations concerned with the issue of environmental justice.

Pollack, Stephanie, and JoAnn Grozuczak. *Reagan, Toxics, and Minorities: A Policy Report by the Urban Environment Conference, Inc*. Washington, DC: Urban Environment Conference, Inc., September 1984. 57 pp.

In his introduction to this book, Representative John Conyers, Jr., (D-Michigan) points out that this report is the first to focus on the fairness of President Ronald Reagan's health and safety policies, which, he claims, "shift the burden of toxic pollutants in the workplace and the environment to minorities." The book itself consists of six chapters covering issues of lead pollution, solid and hazardous waste management policies, farmworker exposure to pesticides, occupational health and safety, special issues of concern to Native Americans, and general conclusions. Conyers concludes from the case studies and data presented in the book that "the Reagan Administration has continued to display remarkable insensitivity to the health concerns of minorities."

The authors of the book come to a similar conclusion. "The blame for the increasing health damage to minorities that will occur for years into the future," they say, "falls collectively to those officials in charge of agencies like EPA, OSHA, and HUD. The policies carried out by these officials reflect those of President Reagan with his ideology of 'ignore the law and blame the victim.'" The personal stories of individual instances of environmental inequities presented in the book are particularly moving.

The Public Health Institute and The Labor Institute. *Jobs and the Environment*. New York: The Council on International and Public Affairs, Inc., Draft 4, May 1994. 190 pp. with appendices and evaluation forms. ISBN 0-945257-62-7.

The Public Health Institute and The Labor Institute worked cooperatively to produce this workbook. It is designed to train and educate "a core of environmental, community justice, and labor activists from around the country who understand each other's concerns and who understand the basic economic forces producing these tensions [between workers and people who struggle for environmental and public health protection]." The workbook consists of nine activities on topics such as "Toxic Trends," "Toxic Roulette," "Job Insecurity and the Labor Market," "The Seamless Workplace: Exposures to Workers and Community," and "Finding Solutions." Each activity consists of background information and data, followed by role-playing "tasks" that ask participants to reach decisions on specific questions and issues.

Sheehan, Helen E. and Richard P. Wedeen, eds. *Toxic Circles: Environmental Hazards from the Workplace into the Community*. New Brunswick, NJ: Rutgers University Press, 1993. 277 pp. ISBN 0-8135-1990-X.

The issue of occupational health has long been, as the editors of this book point out, a stepchild of the medical profession. In fact, it was only recently that the study of occupational health earned respectability as a legitimate profession among physicians.

Yet, problems of occupational health have been around for centuries. In this book, 12 authors discuss case studies in occupational health, all taken from a single state, New Jersey. The cases include the problem of mental disorders among hatters, the incidence of cancer among radium dial painters, the occurrence of bladder cancer among workers in the dye industry, and the prevalence of scrotal cancer among petroleum workers. The book illustrates over and over again how common it is for corporations to knowingly place workers into risky situations.

Smith, James Noel, ed. *Environmental Quality and Social Justice in Urban America*. Washington, DC: The Conservation Foundation, 1974. 145 pp.

This book deals with the core issues of environmental racism and environmental justice at an early period (1974) in the modern environmentalist movement. The editor points out that some observers have raised the question as to whether "the goals and strategies of the environmental movement are somehow antithetical to those interests of society which are seeking social justice and equality of opportunity, and that an elite, upper-middle-class, exclusively white sector of America is using the environmental issue, either overtly or unconsciously, to protect its own 'room at the top' from the encroachments of those less favorably placed on the social and economic ladder of American society." Although the editor feels that such claims may be "far-fetched," he admits that "one cannot be oblivious to the increasing conflict that characterizes such topics as environmental-quality protection and jobs, low- and moderate-income housing, open-space and landscape protection, aesthetic taste versus utilitarian services, and, last but not least, that conundrum of how to make an equitable social distribution of the costs of pollution control."

The volume consists of a series of papers submitted at a three-day seminar on the subject of *Environmental Quality and Social Justice* sponsored by the Conservation Foundation in November 1972. The papers deal with topics such as "Conservation for Whom?," "The Inner City," "Organized Labor," "The Double Standard of Open Space," and "Controlling Economic Growth." The fundamental concepts of environmental racism, environmental inequities, and environmental justice as they are known today are nowhere specifically discussed, but the core issues of environmental quality and social justice run through the papers and discussions as clearly as they do in any book published on these topics in the last few years.

Szasz, Andrew. *EcoPopulism: Toxic Waste and the Movement for Environmental Justice*. Minneapolis: University of Minnesota Press, 1994. 216 pp. ISBN 0-8166-2175-6 (paperback).

Issues of environmental justice are discussed within the broader context of the movement that has developed around the issue of toxic wastes. Szasz argues that more traditional methods of dealing with environmental issues have not been effective in confronting the growing problem of toxic and hazardous wastes and that other approaches, such as action on local and regional rather than national scales, have become more important. The title of the book reflects Szasz's view that the environmental justice movement represents the birth of "a radical environmental populism." This movement has spread beyond the original bounds of toxic and hazardous waste issues to include municipal wastes, military toxics, and pesticide pollution.

Szasz further points out that an essential feature of this new movement is a greater concern about and emphasis on the prevention of environmental problems rather than on efforts to clean them up after they have been created. The book is useful because it examines the role and impact of the environmental justice movement within both the environmentalist and civil rights movements in the United States.

U.S. Environmental Protection Agency. Environmental Equity Workgroup. *Environmental Equity: Reducing Risk for All Communities*. Washington, DC: Environmental Protection Agency, 1992. 2 vols., 43 and 128 pp. Documents EPA230-R-92-008 and EPA230-8-92-008A.

The Environmental Equity Workgroup was appointed by Administrator William K. Reilly in July 1990 to review evidence that racial minority and low-income communities bear a disproportionate environmental risk burden. These two volumes constitute the report of the workgroup. Volume One contains the introduction and executive summary, background and context, findings, recommendations, and descriptions of existing EPA projects. Volume Two contains supporting documents used in the preparation of Volume One. It includes information on health and exposure data; evaluation of programs operated by the EPA; issues of special concern to Native Americans; findings on risk assessment, risk management, and risk communication; outreach efforts; regional perspectives; an institutional model for dealing with environmental equity issues; and comments from external reviewers.

U.S. General Accounting Office. *Siting of Hazardous Waste Landfills and Their Correlation with Racial and Economic Status of Surrounding Communities*. Washington, DC: Government Printing Office, 1983. 20 pp.

One of the landmark studies in the early history of the environmental justice movement was conducted by the U.S. General Accounting Office (GAO). Following the protest in Warren County, North Carolina, regarding the siting of a PCB (polychlorinated biphenyl) landfill, Representative Walter Fauntroy (D-Washington, D.C.) asked the GAO to "determine the correlation between the location of hazardous waste landfills and the racial and economic status of the surrounding communities." This report presents the results of the GAO study.

The GAO found that there were four off-site hazardous waste landfills in the eight states that make up the EPA's Region IV. African Americans made up the majority of the population in three of the four communities in which the landfills were located. In addition, the GAO found that at least 26 percent of the population in all four communities had income less than that of the poverty level, and the majority of that population was African American.

Although the GAO report is relatively brief (only 20 pages), the research findings are presented clearly and starkly in eight table and maps showing the disproportionate distribution of landfills in communities of color and low income.

Vallette, Jim. *The International Trade in Wastes: Policy and Data Analysis by Greenpeace International*, 2nd edition. Washington, DC: Greenpeace, 1988. 69 pp.

Greenpeace has for many years conducted a study of the trade in hazardous and nonhazardous wastes taking place between various countries of the world. This publication reviews accepted international policy on the transfer of wastes, offers five major recommendations on this topic, and then summarizes the group's research on specific waste shipments (classified by inventory number). The report was reprinted in its entirety in

International Export of U.S. Waste, hearings before a subcommittee of the committee on government operations of the House of Representatives, 100th Congress, Second Session, 14 July 1988, pp. 381–499.

Wasserstrom, Robert F., and Richard Wiles. *Field Duty: U.S. Farmworkers and Pesticide Safety.* Washington, DC: World Resources Institute, 1985. 78 pp. ISBN 0-915825-08-2.

Although now somewhat dated, this book is an excellent resource on issues concerning the exposure of farmworkers to pesticides. It is widely quoted in works on environmental justice.

Commissioned by the World Resources Institute, the study covers the history of farmworker exposure issues and existing research on the subject, concluding with recommendations for dealing with farmworker exposure in the future. Some topics included in the book are case studies of farmworker exposure to pesticides; the basis for and development of federal and state regulations on farmworker exposure; scientific evidence on various pesticide properties, such as rate of degradation; problems with the application and enforcement of regulations; training requirements and programs for pesticide applicators; the effects of changing national administrations on pesticide use policies; the efficacy of various types of protection for farmworkers; and data on accidents resulting from the use of pesticides in agriculture.

Weinberg, Bill. *War on the Land: Ecology and Politics in Central America.* London: Zed Books, 1991. 203 pp. ISBN 0-86232-946-9.

In the preface to this book, the author explains that he "came of age politically" in the anti-nuclear movement of the late 1970s and early 1980s. One of the lessons he learned in that movement, he explains, was the confluence of two powerful social movements, the leftist social movement and the environmental movement. When the anti-nuclear movement dissipated during the Reagan era, he goes on, he felt that there must be other areas in which these two otherwise diverging movements should find common ground. That common ground, it turned out, was in the ecological destruction taking place in many parts of the world, such as Central America.

The book is divided into three parts. The first outlines some of the forces that have led to ecological assaults in Third World nations—forces such as cotton cultivation and beef growing and exportation of wastes to poor nations. The second deals with issues faced in each of the seven Central American nations, ranging from the use of valuable land in Guatemala for cattle ranching to the ecological effects of militarization and civil wars in El Salvador, Honduras, and Nicaragua. In the third, the author suggests some possible solutions to the ecological crises faced in Central America.

Wenz, Peter S. *Environmental Justice*. Albany: State University of New York Press, 1988. 368 pp. ISBN 0-88706-644-5.

The title of this book should not mislead a reader to believe that it deals with the environmental justice movement that is the subject of this book. Instead, it concerns more generally the question of how benefits and burdens are to be distributed in a society. The author points out in the introduction that the Arab oil embargo of 1973–74 raised a number of questions regarding how scarce but valuable resources were to be allocated. He concluded that in the debate over changes that might be necessitated by that embargo, "the environment would be spared some damage, but people would not be. Poorer people would be hurt, and, it seemed to me, justice would suffer as wealth was transferred from the relatively poor to the relatively affluent." Thus, although not addressed to issues of environmental justice in the context presented in this book, Wenz's work does deal quite directly with many of the fundamental philosophical issues embraced by that movement. Interestingly enough, he points out (as do some activists in the environmental justice movement) that his own work "is devoted more systematically to the topic of justice than to that of the environment."

Zupan, Jeffrey M. *The Distribution of Air Quality in the New York Region*. Baltimore, MD: Johns Hopkins University Press for Resources for the Future, 1973. 87 pp. ISBN 0-8018-1540-1 (paperback).

This study is an early attempt to discover how the costs and benefits of an environmental problem—air pollution—are distributed across income groups. The author studied this question in the 31-county, 3-state region surrounding New York City. He analyzed exposure to a variety of pollutants, such as sulfur dioxide, particulates, and other residuals. The results obtained in the study are not easy to summarize briefly and vary depending on the measure of pollution used. But one finding is that, at least in some cases, relatively small differences in exposure to air pollution existed across income groups except for those in the very lowest income group, where exposure was most serious and benefits gained from air quality laws were least effective.

Reports

Basel Convention on the Export of Waste. Hearing before the Subcommittee on Transportation and Hazardous Materials of the House Committee on Energy and Commerce, 102nd Congress, 1st Session, 10 October 1991.

Disproportionate Impact of Lead Poisoning on Minority Communities. Hearings before the Subcommittee on Health and the Environment of the House Committee on Energy and Commerce, 102nd Congress, 2nd Session, 25 February 1992.

Environmental Issues. Hearings before the Subcommittee on Transportation and Hazardous Materials of the House Committee on Energy and Commerce, 103rd Congress, 1st Session, 17 and 18 November 1993.

Environmental Justice. Hearings before the Subcommittee on Civil and Constitutional Rights of the House Committee on the Judiciary, 103rd Congress, 1st Session, 3 and 4 March 1993.

Environmental Justice 1994 Annual Report: Focusing on Environmental Protection for All People. Publication EPA-200-R-95-003 of the Environmental Protection Agency, April 1995.

Environmental Justice Strategy: Executive Order 12898. Publication EPA-200-R-95-002 of the Environmental Protection Agency, 3 April 1995.

Environmental Protection Agency Cabinet Elevation—Environmental Equity Issues. Hearing before the Legislation and National Security Subcommittee of the House Committee on Government Operations, 103rd Congress, 1st Session, 28 April 1993.

International Export of U.S. Waste. Hearing before a Subcommittee of the House Committee on Government Operations, 100th Congress, 2nd Session, 14 July 1988.

Selected Nonprint Resources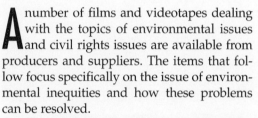

A number of films and videotapes dealing with the topics of environmental issues and civil rights issues are available from producers and suppliers. The items that follow focus specifically on the issue of environmental inequities and how these problems can be resolved.

The reader is also referred to videotapes dealing with aspects of occupational hazards available from The Labor Institute, 853 Broadway, Room 2014, New York, NY 10003 (tel: (212) 674-3322). Included in this series are titles such as "Multiple Chemical Sensitivities at Work," "Surrounded: the Occupational Health Hazards of EMFs," and "More Than We Can Bear: Reproductive Hazards on the Job."

Another source of visual aids dealing with workplace hazards is the Labor Occupational Health Program at the University of California at Berkeley's School of Public Health, 2515 Channing Way, Berkeley, CA 94720-5120 (tel: (510) 642-5507 or fax: (510) 643-5698). Some of the titles available from LOHP include "Preventing Terminal Illness," "It Didn't Have To Happen," "Working for Your Life," "Lead in Construction Training," and "Esto No Tenia Que Pasar."

Deadly Deception
Type: Videocassette
Age: High school to adult
Length: 29 min.
Cost: Institutions: sale, $50; rental, $25
Individuals/low-income groups: sale, $25
Date: 1991
Source: The Video Project
5332 College Ave., Suite 101
Oakland, CA 94618
(510) 655-9050; (800) 4-PLANET
(510) 655-9115 (FAX)
videoproject@igc.apc.org (e-mail)
http://www.videoproject.org/videoproject/
(web address)

The 1991 Oscar for Best Short Documentary went to this film. It describes how wastes produced by several nuclear weapons plants operated by the General Electric Company caused a variety of health problems in neighborhoods surrounding the plants. The film "raises significant questions about corporation responsibility and the role of citizens in challenging abusive business practices."

Deafsmith, A Nuclear Folktale
Type: Videocassette
Age: High school to adult
Length: 43 min.
Cost: Institutions: sale, $79; rental, $45
Individuals/low-income groups: sale, $39.95; rental, $25
Date: 1990
Source: The Video Project (see first entry)

In the late 1980s, the U.S. Department of Energy (DOE) proposed constructing a permanent burial site for the nation's high-level radioactive wastes in Deafsmith County, Texas. This video portrays the battle of county residents to keep DOE from following through on its plans. Representatives from DOE and county residents tell their sides of the controversy, resulting in "a good overview of the problems posed by nuclear waste disposal."

Eco-Rap: Voices from the Hood
Type: Videocassette
Age: 9 and up

Length: 38 min.
Cost: Institutions: sale, $59.95; rental, $30
         Individuals/low-income groups: sale, $29.95; rental, $20
Date: 1993
Source: The Video Project (see first entry)

Environmental experts lead a multiethnic group of young men and women on a tour of urban toxic sites, including hazardous waste sites and other sources of pollution, pointing out that urban residents are faced with a whole set of their own environmental issues. The young men and women develop a series of rap songs to tell about what they have learned from the environmentalists. The best of the songs are then selected for presentation at an eco-rap concert in a local city park. A study/action guide is included with the video.

From the Mountains to the Maquiladoras:
A TIRN Educational Video
Type: Videocassette
Age: High school to adult
Length: 25 min.
Cost: Rental: $25
Date: 1993
Source: Highlander Center
         1959 Highlander Way
         New Market, TN 37820
         (423) 933-3443
         (423) 933-3424 (FAX)

In 1991, the Tennessee Industrial Renewal Network (TIRN) sponsored a Worker Exchange Program with members of the Committee Fronterizo de Obreras (Border Committee of Women Workers) near Matamoras, Mexico. The women were workers in *maquiladora* factories located near the U.S.-Mexican border. This video includes discussions between members of the two groups, along with visits to the "colonia" in which the Mexican women work and an interview with the manager of a General Motors plant in the area. It "provides a look at the reality of life in the Maquiladora region while demonstrating the importance of worker-to-worker exchanges in our increasingly global economy."

Get It Together
Type: Videocassette
Age: 12 and up

Length: 28 min.
Cost: Institutions: sale, $59.95; rental, $30
Individuals/low-income groups: sale, $29.95; rental, $20
Date: 1993
Source: The Video Project (see first entry)

Produced by young people for young people, this video illustrates the impact that children and young adults can make on environmental problems in their own communities. Some of the organizations profiled in the video are Detroit Summer, a group working to improve America's inner cities; YouthBuild, a New York group that constructs low-income housing; and Mexican-American women activists from California working on problems of pesticide exposure. The video is intended to "empower youth" and show them how involvement can "help them develop a sense of pride in themselves and their communities." A study guide is included with the video.

Global Dumping Ground
Type: Videocassette
Age: High school to adult
Length: 58 min.
Cost: Sale, $39.95; rental, $20
Individuals/low-income groups: sale, $39.95
Date: 1990
Source: The Video Project (see first entry)

Originally produced as a PBS Frameline special, this video describes the problems created by exporting wastes from developed nations of the world to the developing nations. The video traces the path of the Colbert brothers, notorious for shipping hazardous wastes in falsely labeled barrels to more than 100 different countries. The video is accompanied by a 152-page book. (See chapter 6 for description of the book.)

Greenbucks: The Challenge of Sustainable Development
Type: Videocassette
Age: High school to adult
Length: 55 min.
Cost: Institutions: sale, $150; rental, $75
Individuals/low-income groups: sale, $39.95; rental, $20
Date: 1992
Source: The Video Project (see first entry)

Many of those interested in issues of environmental inequities are beginning to recognize that a critical element in the solution of such problems is the need for greater attention to reducing environmental problems of all kinds, rather than trying to clean them up after they have occurred. The concept of sustainable development—development that uses natural resources more wisely—is thus becoming more closely linked with that of environmental justice. In this video, executives from some of the world's largest corporations discuss how they are changing the way they operate to meet the goals of sustainable development. The program was first aired on the BBC in 1992 and was awarded the Silver Apple at the National Educational Film & Video Festival that year.

Harvest of Sorrow: Farm Workers and Pesticides
Type: Videocassette
Group: Farmworkers
Length: 30 min.
Cost: See below
Date: 1988
Source: Pesticide Education Center
          P. O. Box 420870
          San Francisco, CA 94142-0870
          (415) 391-8511
          (415) 391-9159 (FAX)

This series consists of two videos and training manuals, available in both English and Spanish, designed to teach farmworkers about the dangers and health effects of pesticides. Part I, "Fieldworkers," describes how pesticides enter the body, the symptoms of poisoning, ways of minimizing exposure to pesticides, and interviews with farmworkers and their families, union leaders, physicians, and an attorney.

The video alone sells for $20.00 and the 70-page training manual, for $10.00. The set of video and manual sells for $25.00.

Part II of the set, "Mixers/Loaders/Applicators," discusses EPA toxicity categories, pesticide labeling and other sources of information available to workers, proper protective clothing and equipment, and unsafe working conditions. Farmworkers from California, Texas, and Florida are also interviewed in the video. The video alone sells for $30.00. The 144-page training manual sells for $15.00. The set of video and manual sells for $40.00.

Heroes of the Earth
Type: Videocassette
Age: 13 to adult
Length: 45 min.
Cost: Institutions: sale, $59.95; rental, $35
Individuals/low-income groups: sale, $39.95; rental, $25
Date: 1993
Source: The Video Project (see first entry)

The seven Goldman Environmental Prize winners for 1993 are profiled in this video, which illustrates the difference that devoted individuals can make in dealing with environmental issues. Among those portrayed is a Native American woman from South Dakota who fought to protect tribal lands from use as hazardous and radioactive waste sites. Study guide included.

Living with Lead
Type: Videocassette
Age: 16 to adult
Length: 58 min.
Cost: Institutions, $79; rental, $40
Individuals/low-income groups: sale, $39.95; rental, $20
Date: 1995
Source: The Video Project (see first entry)

This video tape consists of two parts, each of which may be shown individually. The first part discusses reasons that lead paint poses an environmental hazard, portrays families that have had to deal with this problem, and describes how they have handled the problem. The second part of the program illustrates how some communities have developed successful lead abatement and prevention programs.

The Moon's Prayer: Wisdom of the Ages
Type: Videocassette
Age: High school to adult
Length: 51 min.
Cost: Institutions: sale, $85; rental, $35
Individuals/low-income groups: sale, $39.95; rental, $20
Date: 1991
Source: The Video Project (see first entry)

Native Americans in many parts of the country are battling to preserve their traditional culture in the face of the white man's civilization encroaching upon them in so many ways. This video shows how tribes in the Pacific Northwest are battling to protect and restore their native lands. Their work is "having a positive impact on environmental policies in their scenic region, and are teaching everyone a great deal about resource conservation."

The River that Harms
Type: Videocassette
Age: High school to adult
Length: 45 min.
Cost: Sale, $39.95; rental, $25
Date: 1987
Source: The Video Project (see first entry)

This video documents the pollution of the Puerco River in New Mexico by a 94-million-gallon release of radioactively contaminated water from the United Nuclear Corporation storage dam in 1979 (see chapter 1). Damage caused by the spill to Navajo ranchers who live along the Puerco is documented and illustrated in the film. The film was awarded a Special Jury Award at the Houston International Film Festival and the Cindy Award for Special Achievement by the Association of Visual Communication.

Toxic Racism
Type: Videocassette
Age: 15 to adult
Length: 56 min.
Cost: Institutions: sale, $149; rental, $49.95
Low-income groups: sale, $39.95; rental, $20
Date: 1994
Source: The Video Project (see first entry)

The video that speaks perhaps most directly of all to the issues of environmental racism in the United States, this program first appeared as a PBS special hosted by award-winning journalist Ira Flatow. The video highlights stories from three communities dealing with problems of environmental inequities: Kettleman City, California, where Hispanic Americans battle the construction of a hazardous waste incinerator; North Carolina, where a combination of African American and white residents challenge the spread of waste-producing hog farms; and West Dallas, Texas, where African Americans struggle to get the government to clean up lead pollution produced in smelting operations. Featured on the video are Robert Bullard and Benjamin Chavis, both activists in the environmental justice movement.

The Yes! Tour: Working for Change
Type: Videocassette
Age: 12 and up
Length: 29 min.
Cost: Institutions: sale, $59.95; rental, $30
Individuals/low-income groups: sale, $29.95; rental, $20
Date: 1992
Source: The Video Project (see first entry)

YES (Youth for Environmental Sanity) is a group of concerned teenagers who travel across the country in an old station wagon talking to students about ways they can work on environmental problems in their own neighborhoods. The video is designed as a mechanism young people can use to begin thinking about environmental issues in their communities and possible ways of dealing with those issues.

Glossary

acceptable risk Possible hazard whose magnitude is considered to be small enough to accept in exchange for some valuable outcome.

accountability An obligation or willingness to accept responsibility for the consequences of one's actions.

air pollution A state in which air contains one or more substances (such as sulfur dioxide) or conditions (such as heat) that may cause harm to plants, animals, or the physical environment.

ambient An adjective referring to surrounding conditions. Ambient air, for example, is the surrounding air.

apartheid A policy of segregating nonwhites from whites that grew up in South Africa and was legitimized by legal and political regulations. Although formal legal strictures in the United States do not exist as they long did in South Africa, some activities use the word to describe the *de facto* (rather than *de jure*) racial discrimination that exists in the United States.

aquifer Permeable layers of underground rock that hold groundwater and through which the groundwater may move. The top layer of an aquifer is known as the water table. Many communities, especially in the western states, obtain their drinking water from aquifers. These communities face serious problems when an aquifer is depleted for or polluted by industrial uses.

BANANA (building absolutely nothing anywhere near anything) principle A term ascribed to some environmentalists who would oppose the construction of any environmentally harmful facility anywhere because of its potential effects on plant, animal, and human life. *See also* **NIMBY** and **NOPE.**

barrio   A district within a Spanish-speaking city or town.

benign neglect   A policy of ignoring an unpleasant situation rather than finding ways to deal with the situation.

best available technology (BAT)   The most efficient and/or effective way of dealing with some environmental issue, such as the control of air pollution, no matter what the cost of that technology may be.

biodiversity   The range of plant and animal life found within an area. Under most circumstances, a healthy ecosystem is one in which there is significant biodiversity. The loss of plant and animal species results in a decrease in biodiversity and a consequent decline in the general health of the ecosystem.

birth defect   A medical disorder that shows up at the time a baby is born. It may be caused by a number of factors, many of which are genetic in origin.

Black Congressional Caucus   A group of Congressmen and Congresswomen of African American heritage who work together on issues of common concern.

blaming the victim   A philosophy that bad things happen to people because it's their own fault. For example, one might argue that African Americans are more likely to have accidents in a particular occupation because they are lazy, careless, or less intelligent when, in fact, another explanation might be that they have been placed into dangerous working conditions to which non-African Americans are not exposed.

blue-collar workers   Wage earners (in contrast to salaried employees), many of whom work in situations in which some form of protective clothing is required.

boomerang effect   An effect that may occur when pesticides or other chemicals banned in the United States are shipped to other nations and then return to this country causing unanticipated health effects. For example, a pesticide banned in the United States because it is too toxic for consumers may still be shipped to Mexico and used by farmers there. The pesticide may then remain on products shipped from Mexico back to the United States, resulting in the health threat that banning in the first place was supposed to prevent. The boomerang effect is also known as the circle of poison.

boycott   A nonviolent form of protest in which a group of people jointly agree not to purchase a product or patronize an establishment in order to demonstrate their opposition to some policy or action supported by the manufacturer or establishment.

buffer zone   A strip of land that separates two other pieces of land from each other. For example, some of the effects of an industrial facility can be ameliorated by installing a buffer zone between it and surrounding neighborhoods.

burden of proof In law, the determination as to who it is that must prove that harm has or has not, will or will not, be caused by some practice. For example, in most cases of environmental inequities now, those individuals or groups who feel that they are being harmed by some environmental nuisance must demonstrate that such harm has actually occurred. Minority and poor communities may not have the resources, experience, and finances to demonstrate such proof. Alternatively, it would be possible to place the burden of proof on the other side of the argument by insisting that industries, for example, demonstrate in advance of constructing a facility that that facility will not constitute a nuisance for surrounding neighborhoods.

business-as-usual The tendency of a company or the government to continue conducting business as it has in the past without regard to conditions that might reasonably warrant changes in that business pattern. For example, an industry that learns that its smokestack emissions are causing health problems in areas surrounding its plants might decide not to do anything about the problem and to continue doing "business-as-usual."

buyout To completely purchase a home, business, or other operation. Companies desiring to expand their operations in surrounding areas, for example, have been known to purchase whole neighborhoods or even towns in order to obtain the land they need for expansion. In other cases, polluting industries have bought out communities as a way of dealing with complaints about the health hazards they present to the area.

Cancer Alley A term used to describe an 80-mile strip of the lower Mississippi River along which more than 100 oil refineries and petrochemical plants are located. According to the U.S. Environmental Protection Agency, the region shows an "unusually high incidence of cancer, asthma, hypertension, strokes, and other illnesses," that may very well be traced to the presence of toxic pollutants in the area.

capability analysis A study of the ways in which a piece of land can be used. Capability analysis stands in contrast to suitability analysis since the former determines what *can* be done and the latter, what *should* be done. *See also* **suitability analysis**.

carbamate A member of a class of chemicals widely used as pesticides. The chemicals act as nerve poisons, incapacitating organisms exposed to them. Sevin and Zireb are examples of carbamate pesticides.

carcinogen Environmental agents capable of causing cancer. Some typical carcinogens are X-rays, ultraviolet light, compounds of nickel and chromium, vinyl chloride, and the tar in cigarette smoke.

chlorinated hydrocarbon A member of a class of chemicals widely used as pesticides. Like the carbamates, they act as nerve poisons. Some well-known examples of the chlorinated hydrocarbons are DDT, aldrin, dieldrin, and chlordane.

chronic bronchitis A common respiratory disorder characterized by inflammation of the bronchial passages, excessive secretions of mucus, and recurring bouts of coughing. Bronchitis is aggravated by and may be caused by exposure to air pollutants.

circle of poison *See* **boomerang effect.**

civil disobedience The refusal to obey certain laws and regulations in order to protest governmental or other policies. The refusal generally takes the form of some nonviolent action, such as lying down in front of oncoming traffic. In most cases, civil disobedience also involves the joint cooperation of many individuals in an action.

civil rights The rights of personal liberty granted to an American citizen through Amendments Thirteen and Fourteen of the U.S. Constitution as well as through other laws and regulations.

class action suit A legal action taken on behalf of all members of a group directly affected by a case. For example, The Sierra Club Legal Defense Fund has undertaken a number of class action suits on behalf of members of a community who have been exposed to a disproportionate share of environmental pollution.

clawback agreement A provision by which one member of an agreement is able to collect compensation, usually in the form of money, from a second party to the agreement if and when that second party does not fulfill its obligations according to the terms of that agreement. The provision has been suggested for situations in which a corporation agrees to provide certain amenities to a community, such as jobs or material improvements in infrastructure, in return for the siting of an undesirable facility. Should the company not follow through on its agreement, the clawback provision would allow the community to recover damages from the corporation.

cleanup The actions taken in order to neutralize and/or remove hazardous wastes from an area.

communities of color *See* **people of color.**

community buyout The practice by which a governmental body or private industry pays for the complete removal of a community because of its proximity to one or more environmentally dangerous facilities. For example, in 1983, the federal government found that more than 100 sites in the state of Missouri were seriously contaminated with the highly toxic chemical known as dioxin. As part of its cleanup effort, it eventually paid for the closing and removal of all residents of the town of Times Beach.

community information statement A feature proposed in Congresswoman Cardiss Collins' bill to amend the Resource Conservation and Recovery Act that would require a study of the demographic characteristics of a community in which a hazardous waste landfill is planned and an assessment of the potential impact of the landfill on the community.

community right-to-know *See* **right-to-know.**

correlation A statistical term that refers to the extent to which two variables are mathematically related to each other. For example, one might discover in a research study that high levels of lead in the blood are mathematically correlated with poor learning skills. A high correlation, however, does not prove cause-and-effect relationships. In this example, the cause-and-effect relationship between blood level and learning has been confirmed by other types of studies than correlational studies.

cost/benefit analysis Any attempt to compare the advantages of taking some action (such as installing a waste disposal incinerator) with the disadvantages of that action (such as the release of harmful gases from the incinerator).

cost effectiveness analysis An attempt to compare the effectiveness and desirability of achieving an improvement in environmental conditions (such as a reduction in air pollution) by various technologies.

DAD (decide, announce, defend) An acronym used to describe a common industrial practice of making decisions about the siting of environmentally harmful facilities. In this practice, the industry makes essentially all relevant decisions about the siting without input from outside groups, such as individuals or groups who might live in the neighborhood.

de facto Latin for "by fact," referring to practices and policies that actually exist, whether or not they are established and maintained by laws and regulations.

de jure Latin for "by right," referring to practices and policies that are established and maintained by law and regulation.

debt-for-nature swap A mechanism by which poor countries agree to set aside parts of their land for conservation purposes in exchange for the cancellation of all or part of a financial debt that they owe to another country or to an international bank. Although debt-for-nature swaps can be an excellent way of saving lands that might otherwise be developed for commercial, agricultural or industrial purposes, the practice tends to interfere with the efforts by indigenous peoples to determine what happens to their own lands.

degradable Capable of decomposing and changing into some other form, often less hazardous or toxic than the original form. A common feature of synthetic materials produced by industry today (such as many forms of plastic) is that they are not naturally degradable and, thus, remain in the environment for hundreds or thousands of years. *See also* **persistent**.

demography The statistical study of human populations, including such information as numbers, distributions among race and ethnic groups, and birth and death rates.

development rights Concessions that allow the conversion of land from natural area or agricultural use to residential, commercial, or industrial use. Development rights can be bought and sold just as can mineral rights or air rights.

discrimination The practice of judging people on the basis of one or more classes to which they belong rather than on the basis of their individual characteristics.

discriminatory intent A legal term that means that plaintiffs in an environmental justice (or other) case must be able to prove that race is a "motivating factor" in a decision made by an individual or corporation that brings harm to a person or group of people. The term arises out of a U.S. Supreme Court ruling in 1976, *Washington v. Davis* (426 U.S. 229, 238–248).

disease of adaptation A term that has been coined to describe a host of medical problems developed by people forced to live in unhealthy environments that include polluted air and water and stressful living conditions.

disproportionate An adjective that refers to the fact that something is made available, distributed, or present to people in a ratio that differs from the percentage of those people in the general population. For example, if African Americans make up 10 percent of the U.S. population and live in neighborhoods where 20 percent of the nation's pollution is produced, then the exposure of African Americans to pollution is disproportionate to their numbers in the general population.

easement A right granted by a landowner to another person or organization to use a piece of her or his land.

ecocide The destruction of a whole biological community. In many cases, ecocide occurs because a relatively small number of organisms are destroyed, resulting in the disappearance of many other organisms intimately associated with those that are wiped out. The term generally carries with it a connotation that human activities are responsible for this process.

ecology The study of the relationships of living organisms with each other and with the physical surroundings in which they are found.

economic blackmail A practice in which a corporation offers financial benefits to individuals or a community in exchange for putting up with environmental or other hazards. Workers might be paid relatively high wages, for example, in return for accepting employment in facilities or jobs that are hazardous to their health.

effluent Any material, usually a liquid, discharged from a point source into the surrounding environment.

emission standard The maximum amount of any pollutant that is permitted by law or regulation to be released into the environment. Emission standards have now been set for the great majority of pollutants, such as sulfur dioxide and carbon monoxide.

encroachment Any entry into a restricted area, usually by illegal means. The construction of a factory or landfill that extends onto someone else's property is an example of encroachment.

environment The sum total of all the living and nonliving conditions that influence the life of an individual organism or a population of organisms.

environmental discrimination The disproportionate exposure to adverse environmental conditions as a result of racial, ethnic, economic, or other characteristics of a community.

environmental equity   A condition in which the burdens and benefits resulting from technological development are shared equally by all groups within society.

environmental high-impact area   An area that is subject to a higher-than-normal concentration of hazardous conditions, such as air and water pollution and/or hazardous waste sites.

environmental impact statement   A document that describes the effects that a proposed action is likely to have on the environment. An environmental impact statement is now required by law before many projects can be initiated. *See also* **impact analysis**.

environmental justice   The attempt to achieve environmental equity for all groups within society.

environmental racism   A term coined in 1982 by Benjamin F. Chavis, Executive Director of the National Association for the Advancement of Colored People. In testimony before Congress, Chavis has said that the term refers to "racial discrimination in environmental policy making and the unequal enforcement of environmental laws and regulations. It is the deliberate targeting," he goes on, "of people of color communities for toxic waste facilities and the official sanctioning of a life-threatening presence of poisons and pollutants in people of color communities. It is also manifested in the history of excluding people of color from the leadership of the environmental movement."

environmental screening   Tests performed to determine the exposure that an individual or a community has had to a hazardous substance. Tests are available, for example, to determine the amount of radon found in a home. Testing blood for the presence of lead is another example of environmental screening.

environmentally disadvantaged community   An area in which there exists at least one hazardous waste facility and that contains a higher-than-average percentage of low-income or minority residents.

environmentally sound management   A term introduced in the Basel Convention on the Control of Transboundary Movements of Hazardous Wastes and Their Disposal to describe an environmentally correct way for handling hazardous waste. The term was not specifically defined in the Convention, nor has it been since then.

ethnic   A term that refers to a minority group, particularly one with a common language and cultural heritage. In this respect, Italians constitute an ethnic group.

executive order   An order issued by the president, a governor, or some other executive order that has the force of law.

external cost   *See* **negative externality**.

fair market value   An honest price for a piece of land or some other piece of property. Fair market value is determined as a result of the barter that occurs between buyer and seller. The presence of an environmental insult near a piece of property (such as a leaking dump near a private home) can have dramatic effects on the fair market value of that property.

fair share legislation A type of legislation that requires that environmental hazards be shared equally within a community and across communities.

feedlot A confined area in which large numbers of livestock animals are enclosed while being fattened for market. Huge amounts of waste matter from the animals accumulate in a feedlot and, if not removed in a timely manner, produce a variety of air, water, and ground pollution in the surrounding environment.

fossil fuel A fuel formed by the anaerobic decay of plant and animal matter. The term usually applies to coal, oil, and natural gas.

genetic damage Damage caused to an organism's genetic makeup, often resulting in death or disease for that organism. A number of factors can cause genetic damage, such as forms of energy (gamma rays and X-rays) and a variety of chemicals (such as compounds of lead and mercury). *See also* **mutagen**.

gentrification The improvement that takes place when a blighted area, such as a run-down inner city neighborhood, is restored. Many communities have been gentrified by their own residents who are no longer willing to accept the decaying condition of their physical environment.

grandfather rights An exemption from a law or regulation because of the fact that a condition existed *before* the law or regulation was adopted. For example, a person might be allowed to remain in the family home even after the property on which it is located has been rezoned for commercial or industrial use.

grassroots A term describing any organization or movement that is derived from the most fundamental level of society; being organized from the bottom up, rather than from the top down.

green revolution A term used to describe the introduction of modern agricultural techniques into less-developed countries. These techniques include the use of new plant varieties, pesticides, chemical fertilizers, and heavy farm machinery.

groundwater Water that sinks into the ground. That water is typically stored in the soil for long periods of time, during which it may slowly flow from one region to another.

growth management All processes involved in controlling the rate and circumstances under which new developments take place in a community.

hazardous waste Any solid, liquid, or gas released into the environment by an industrial process or from a municipal outlet that may cause damage to the health of a living organism.

heavy metal An element with a relatively high atomic number, such as mercury, cadmium, lead, zinc, or nickel. Many heavy metals have toxic effects on living organisms. They are commonly released in the wastes of industrial processes or as components of municipal wastes.

herbicide   A chemical used to kill weeds or other undesirable plants.

Hispanic American   An American citizen whose ancestry can be traced to a South or Central American nation. *See also* **Latino(a).**

historic preservation   The protection of buildings, districts, objects, and structures that have historical significance. People can sometimes use state and federal historic preservation laws to protect their communities from destruction by industries or other environmentally undesirable developments.

historically black colleges and universities (HBCU)   Colleges and universities that were usually founded as and for many years continued to operate as segregated institutions. Such institutions are now integrated, but continue to play an important educational and research role in the black community.

host fee   A payment made to individual homeowners or to the general fund of a community in return for its accepting the siting of a polluting or hazardous facility in its area. *See also* **offsetting benefit.**

hot potato   A term coined by Ruth Norris to describe a practice that sometimes occurs in developing nations when farmers are unable to purchase the expensive nonpersistent pesticides that will allow their products to be exported to the United States and other developed nations. When that happens, farmers are likely to sell their products, containing dangerously high levels of more persistent pesticides, in their own country.

hot spot   An area that contains a dangerously high level of some hazardous material.

impact analysis   A study conducted to find out how development will affect a particular area. *See also* **environmental impact statement.**

incineration   The process of burning. Incineration is an alternative method of disposing of hazardous wastes other than burying them in a landfill.

indigenous people   People who are native to a particular area, region, or country. Native Americans are indigenous to the United States.

industrial smog   Air pollution caused primarily by the combustion of fossil fuels in industrial plants, commercial buildings, and residences.

insecticide   Any synthetic or naturally occurring substance that kills or disables an insect or prevents the insect from reproducing. *See also* **pesticide.**

institutional racism   A form of racism that is maintained by a variety of legal and customary practices. Segregated schools are an example of institutional racism.

integrated pest management (IPM)   An agricultural concept and practice in which pests are controlled by some combination of natural, biological, chemical, cultural, and physical processes.

internal costs   Costs of production that are paid by the producer or the consumer.

job blackmail   *See* **economic blackmail.**

laissez-faire   From the French, the principle of allowing people to act as they please. In terms of land use, laissez-faire means that landowners can do whatever they want with their own property, and that government should not interfere with the decisions made by private citizens or corporations.

land use   The use to which a piece of land is put. Some of the most difficult issues in modern society concern competing demands for land utilization among industry, recreation, private housing, commercial development, and other possible uses.

landfill   A waste disposal site. The simplest type of landfill is a dump in which materials are discarded with little or no attempt to contain or control the escape of harmful substances into the soil, water, or air. A sanitary landfill is one in which such efforts have been made as, for example, by burying wastes under layers of dirt or by sealing the bottom of the landfill with plastic or some other nonporous material.

Latino(a)   A person of Latin American origin or of Latin American heritage. *See also* **Hispanic American.**

LD-50 (lethal dosage-50 percent)   The amount of a substance that results in the death of one half of the individuals exposed to the substance.

less-developed countries (LDC)   Nations that tend to have low per capita average income, a high rate of population growth, an agricultural rather than industrial economy, and a weak economy and poor financial base.

litigation   Legal action.

lobby   A person or group that tries to influence a legislative or regulatory body.

low-income housing   Residential facilities designed and built for families whose income is less than some standard set by a government agency.

LULUs (locally unwanted land uses)   Facilities such as polluting factories, hazardous waste dumps, and strip mines that make significantly negative environmental contributions to the region in which they are located.

mainstream environmentalism   A form of environmentalism that has grown up over the past century or more in which the emphasis has been on the conservation or preservation of natural resources or the solution of pollution problems. In most cases, mainstream environmentalists have been relatively uninterested in environmental problems of urban areas.

maquiladoras   Industries and plants that operate on the Mexican side of the border between the United States and Mexico, owned by U.S., Japanese, and Mexican companies and employing low-wage Mexican workers.

market forces   The set of conditions that determine the demand for sale of goods and/or services.

Monitored Retrieval Storage (M.R.S.)   A program developed by the U.S. Department of Energy for the temporary storage of spent nuclear fuel rods. The term "temporary" refers to a period of up to about 40 years.

more-developed countries (MDC)   Nations that tend to have a relatively high per capita average income, a low rate of population growth, an industrial rather than agricultural economy, and a strong economy and financial base.

move-to-the-nuisance   A phrase that describes the tendency of people from lower economic groups to move into less desirable living areas (such as those in which pollution is a serious problem) because housing prices are lower in those areas.

mutagen   Any substance or form of energy that can bring about a change in the genetic makeup of an organism.

negative externality   A term used by economists to describe an unpleasant condition (such as the presence of air pollution) that is beyond the control of those people who are exposed to it. The condition is also known as an external cost.

negotiated compensation   A proposed method for dealing with the inequities of environmental racism by charging those entities that create environmental problems (such as the owners of a waste incineration plant) while compensating those who have to live in the area where the environmental degradation has taken place.

networking   The exchange of information or services among individuals and groups with certain common interests. For example, local environmental justice groups often stay in touch with each other through relatively informal means such as state and regional meetings, newsletters, and personal contact through telephone calls, internet communication, and letter-writing.

New Federalism   The policy of shifting many governmental functions from the federal government to state and local governments. This policy came into favor with the election of President Ronald Reagan, who felt that many regulatory functions could be carried out less expensively and more efficiently by local and state governments, rather than by the federal government. A number of critics have felt that the effect of this policy, however, is to greatly weaken a number of regulatory activities, including protection of the environment.

NIMBY (not in my back yard)   A phrase used to describe the opposition of individuals who oppose the siting of any environmentally harmful facility anywhere near the property they own or reside on.

nitrogen oxides   Chemical compounds consisting of nitrogen and oxygen commonly formed during the combustion of fossil fuels, as in electrical power generation plants and in automobile engines.

nonconforming use An activity that was once legal, but that has become illegal as the result of the passage of some new law or regulation.

nongovernmental organization An organization with no association with government at any level. The Sierra Club, the Commission for Racial Justice of the United Church of Christ, and the National Association for the Advancement of Colored People are examples of nongovernmental organizations.

NOPE (no place on Earth) A somewhat pejorative term sometimes used to describe the position of "radical environmentalists" who oppose the construction of any facility that would produce pollution or other effects damaging to the environment anywhere on Earth.

nuclear waste Radioactive materials that remain after they have been used for some purposes, such as fueling a nuclear reactor or arming a nuclear warhead. Such wastes represent a serious environmental hazard since they tend to give off high levels of radiation for very long periods of time (hundreds or thousands of years).

nuisance law Any regulation that prohibits a person or business from interfering with others' enjoyment and use of property. Nuisance laws have a very long history, going back at least to the days of ancient Rome when penalties could be applied to anyone who polluted the city water system.

offsetting benefits Certain advantages given to a community in exchange for its accepting an environmentally hazardous facility in its area. For example, a community may be granted a tax break by the city or state in exchange for accepting the installation of a new hazardous waste facility in its neighborhood. *See also* **host fee**.

open dump A land disposal site in which little or no effort is made to contain wastes within the site. Open dumps are a serious source of air, land, and ground pollution and are now illegal in many parts of the United States. Many dumps no longer in use, however, still exist and have never been cleaned up.

particulate matter Tiny particles of solid matter or tiny droplets of liquid suspended in air. Particulate matter is a major component of polluted air.

people of color Individuals whose skin pigmentation may be other than white (such as brown, black, or yellow), but, more importantly, whose culture is different from that of white Americans from a European culture. African Americans, Hispanic Americans, Native Americans, Asian Americans, and Pacific Island Americans are usually regarded as belonging to communities of color.

persistent Anything that remains in the environment for long periods of time.

pesticide Any substance that kills or disables a pest or that prevents the pest from reproducing. Although the term is commonly used to refer to synthetic chemicals developed by scientists for use against pests, it applies as well to naturally occurring substances with the same effects.

Specific classes of pesticides are named according to the kind of pest against which they work, such as insecticides, rodenticides, fungicides, and nematocides.

pesticide treadmill   A term used to describe the dilemma faced by farmers when they discover that previous levels of a pesticide are no longer sufficient to control the pests for which they are used. When that occurs, farmers must use higher levels of pesticides, resulting in more resistant organisms, resulting in the need for higher levels of pesticides, and so on.

petrochemical   A term referring to the development, processing, sale, and application of chemicals obtained from petroleum or natural gas.

petrochemical colonialism   A term that has been used to describe the practice current in some areas in which the presence of large petrochemical plants establishes a particular kind of economic and political system similar to that in colonial areas or to the pre-Civil War southern system.

pH   A measure of the acidity of a solution. The lower the pH of a solution, the more acidic it is.

photochemical smog   Polluted air produced by the action of light on certain chemicals (such as hydrocarbons and oxides of nitrogen) released by industrial and other processes.

PIBBY (put it in the blacks' backyard)   A policy, usually unspoken, by which facilities with adverse environmental effects are sited in neighborhoods occupied primarily by African Americans or other nonwhite groups.

point source   Any readily identifiable location from which pollution is released. A smokestack or sewage outlet pipe is an example of a point source.

pollution   Any unfavorable change in the environment that may have deleterious effects on plants, animals, or the physical characteristics of the environment.

polychlorinated biphenyls (PCBs)   A class of synthetic chemicals that has a number of important industrial uses, but that is very persistent in the environment with serious health hazards for organisms.

prior informed consent   A term that means that a pesticide that has been banned, withdrawn, or severely restricted in one country can not be exported to another country until and unless the second country has been informed of such action and has agreed to accept the pesticide under these conditions. The policy of prior informed consent was first adopted in 1985 with the adoption of the United Nations' "International Code of Conduct on the Distribution and Use of Pesticides."

proof of intent   A legal term that means that a complainant in a case must be able to show that a person or a company planned or knew that some damage would result from his, her, or its actions.

property value   The value in dollars or some other form of currency of anything that is owned. The property value of land usually refers to the value of that land plus any buildings on the land.

quality of life The extent to which a person is satisfied with his or her life. Some factors that influence quality of life include jobs, schools, transportation, recreational opportunities, and quality of the surrounding environment.

racism Prejudice and/or discrimination based on the belief that some races are inherently more superior to others and that the traits of an individual are entirely or largely determined by the race to which one belongs. *See also* **institutional racism**.

radioactive waste The wastes produced from weapons research, nuclear power plants, medical applications, and other processes in which radioactive materials are used. Radioactive wastes are usually classified either as low-level wastes, which present relatively modest environmental problems, and high-level wastes, which present serious and long-term environmental problems.

rebuttable presumption An assumption about a situation that anyone is allowed to challenge. As used in environmental law, the term usually means that a corporation is presumed to be responsible for some environmental damage, although it is permitted to try to disprove that presumption. The concept is important in environmental equity cases because if not otherwise stated, plaintiffs in a case are required to prove that a corporation or individual is responsible for damages that occur. Under the doctrine of rebuttable presumption, the situation is reversed, and it is up to the corporation or individual to prove that it is *not* responsible for the damage.

recycling The process by which materials are collected, treated (if necessary), and then reused in the production of new products. *See also* **resource recovery.**

Reserved Rights Doctrine A legal principle that states that rights mentioned in treaties between the U.S. government and Indian tribes were not granted by the government, but were rights that belonged to Native Americans in the first place and that were only confirmed by the treaties.

resource recovery The extraction of useful materials and/or energy from wastes. One form of resource recovery is recycling.

right-to-inspect A policy under which industries extend to workers, neighbors, and others the right to enter and inspect their facilities to become more familiar with any potential environmental threat they may pose.

right-to-know A policy under which industries are required to provide information to workers, neighbors, and others exposed to hazardous environmental conditions the nature of those conditions.

risk assessment A study of the harm that may come to a person or community as the result of some action.

runoff Waters that drain off the surface of land into lakes, rivers, streams, oceans, and other bodies of water. Runoff is a serious environmental problem in some locations because such waters dissolve or otherwise carry with them materials, such as pesticides and fertilizers, that may cause harm to plants, animals, and/or the physical environment.

sacrifice zone An area in which there is an unusually high concentration of industries releasing pollutants to the surrounding region.

sanitary landfill A waste disposal site that is constructed to reduce to the greatest extent possible the escape of hazardous, toxic, or other materials into the surrounding environment. As an example, many sanitary landfills have an impermeable underliner that prevents liquids from leaching into the underlying soil. *See also* **secured landfill.**

secured landfill A type of sanitary landfill in which extra precautions are taken to ensure that wastes do not escape into the surrounding environment. For example, wastes may first be sealed in concrete containers or metal drums and then buried in impermeable strata of rocks. The landfill itself may then be monitored continuously to guard against any failure in the security system. *See also* **sanitary landfill.**

self-determination The practice by which the residents of an area are able to make decisions as to how their own lands are to be used. In many underdeveloped countries, indigenous peoples long ago lost the right of self-determination to federal governments and are now trying to regain that right. In developed countries, people of color and/or poor people often have little or no opportunity to participate in decisions as to how the land on which they live is to be utilized.

site To locate.

sit-in A nonviolent form of protest in which individuals simply sit down in an office or some other place of business to prevent the normal transaction of the day's work.

source reduction An approach to dealing with hazardous environmental conditions by reducing the amount of hazardous materials produced during the process of manufacture or production rather than finding ways to clean them up after they have been released to the environment.

Standard Metropolitan Statistical Area (SMSA) As defined by the U.S. Census Bureau, an urban area having a central city (or pair of cities) and suburban area with a population of more than 50,000 residents and an average population of at least 1,000 persons per square mile.

stressors Factors that can cause stress in a person's life. In studies of occupational health problems, for example, a number of working conditions have been found to be stressors, that is, to have contributed to the development of health disorders.

strip mining *See* **surface mining.**

subsidence Sinking of the earth as the result of the removal of underlying rocks and soil. Subsidence is a common problem in coal mining areas where underground mines have collapsed, causing overlying layers of ground to collapse also.

suitability analysis   A study to find out if a piece of land should be used for some given purpose, such as urban development. The term is in contrast to capability analysis, which is a study of the *ability,* not the *desirability,* of carrying out such a development. *See also* **capability analysis**.

sulfur oxides   A generic term used for sulfur dioxide and sulfur trioxide, two gases released during the combustion of fossil fuels as, for example, in electrical power-generating plants or factories.

sunshine laws   Laws that require governmental bodies to hold their meetings in public and to make all official records public. Many states and municipalities now have sunshine laws.

Superfund   A fund established by the Comprehensive Environmental Response, Compensation, and Liability Act to clean up some of the worst hazardous waste sites in the United States. Of the thousands of such sites in the United States, only those designated as the most dangerous are placed on the National Priorities List (NPL), making them eligible for Superfund action.

super-pests   A term sometimes used to describe pests that have become resistant to pesticides formerly used to control them. Super-pests are a problem for a number of reasons. For example, as they began to evolve, larger and larger quantities of pesticides are used to keep them under control, with increasing danger to farmworkers who use the pesticides. Also, super-pests may become so resistant to chemical means of controlling them that farmers are no longer able to keep them under control, they devastate crops, and whole regions may go into economic collapse.

surface mining   The removal of minerals from the Earth's surface by first removing the overlying ground cover on top of the minerals.

sustainable development   Development that takes place with minimal harmful effects on the physical and biological environment.

teratogen   A chemical that, when ingested during pregnancy, causes birth defects.

Third World   A term originally suggested to refer to nations of the world not aligned with either the Communist or non-Communist blocs. Today, it tends to refer to underdeveloped nations of the world, found primarily in Africa, parts of Asia, and South and Central America.

threshold effect   A consequence that occurs only when a certain level of exposure has occurred. For example, a person can inhale very small amounts of carbon monoxide gas without experiencing any problems. But once a certain level of the gas accumulates in one's system, a variety of physiological effects begin to occur.

throw-away community   A term used to describe a community of color or a community of poor people held in so little regard that corporations see no problem in siting hazardous waste dumps, polluting industries, or other LULUs in the community.

tort   A wrongful act other than a breach of contract for which relief is sought through the legal system.

toxic   Poisonous.

toxic colonialism   The practice of exposing people in other countries to environmental hazards that are generally not permitted in industrialized nations such as the United States. The export of hazardous wastes from American cities to African nations is an example of toxic colonialism.

toxin   A poisonous substance produced by a living organism.

variance   Special permission that is granted to someone to deviate from a zoning regulation. As an example, a planning board might grant permission for the construction of a hazardous waste dump in an area that is zoned for residential development only.

(to) vote with one's feet   A phrase used to describe the act of leaving an area in which one no longer wishes to live. In general, middle- and upper-class whites are more able to move out of an area that has become environmentally hazardous or otherwise undesirable as a place of residence. In contrast, people of color, especially poor people of color, are less likely to have this option.

water table   The level beneath the earth's surface below which the ground is saturated with water. The health of the water table is an important issue in many locations because people obtain their drinking water from wells that draw from the water table.

white flight   The tendency of white people—usually of middle or upper class—to move away from areas that become less desirable as residential areas, as when a hazardous waste dump is located in or near an existing residential neighborhood.

zoning   A method of setting aside certain tracts of land for specified purposes. One goal of zoning is to protect the places that people live, go to school and church, and enjoy their recreation, keeping them separate from commercial operations such as factories and waste disposal sites.

Acronyms and Abbreviations

ACLU   American Civil Liberties Union

ATSDR   Agency for Toxic Substances and Disease Registry

CADRE   California Alliance in Defense of Residential Environments

CBA   Citizens for a Better America

CBC   Congressional Black Caucus

CBE   Citizens for a Better Environment

CCHW   Citizens Clearinghouse for Hazardous Waste

CDC Centers for Disease Control and Prevention

CERCLA Comprehensive Environmental Response, Compensation, and Liability Act

CTWO Center for Third World Organizing

CWM Chemical Waste Management, Inc.

DOE U.S. Department of Energy

EIS environmental impact statement

EPA U.S. Environmental Protection Agency

FDA U.S. Food and Drug Administration

FIFRA Federal Insecticide, Fungicide, and Rodenticide Act

FLOC Farm Labor Organizing Committee

GAO U.S. General Accounting Office

GCTO Gulf Coast Tenants Organization

HUD U.S. Department of Housing and Urban Development

LULU locally unwanted land uses

MELA Mothers of East Los Angeles

NAACP National Association for the Advancement of Colored People

NEPA National Environmental Policy Act

NGO nongovernmental organizations

NIEHS National Institute for Environmental Health Sciences

NIMBY not in my backyard

NRDC Natural Resources Defense Council

OSHA Occupational Safety and Health Act

OSMRE U.S. Office of Surface Mining Reclamation and Enforcement

OTA Office of Technology Assessment

PCB polychlorinated biphenyl

PUEBLO People United for a Better Oakland

RCRA   Resource Conservation and Recovery Act

SNEEJ   Southwest Network for Environmental and Economic Justice

SOC   Southern Organizing Committee

USGS   U.S. Geological Survey

WIPP   Waste Isolation Pilot Project

WMI   Waste Management, Inc.

Index

Acceptable risk, defined, 243
Accountability, defined, 243
Achedemade Bator, Ghana, 13–14
Acid Rain Foundation, 17
Agency for Toxic Substances and
    Disease Registry (ATSDR),
    63, 64, 174
Air pollution, defined, 243
  effects on low income groups,
    21
Alabama Freedom Riders of 1961,
    19
Alinsky, Saul, 70
Alsen, Louisiana, 10
Alter, Harvey, 48
Ambient, defined, 243
Americans for Indian
    Opportunity, 77, 174–175, 208
Anderson, Douglas L., 30
Apartheid, defined, 243
Aquifer, defined, 243
Arkansas State Law on
    Environmental Equity in
    Siting High-Impact Solid
    Waste Management Facilities,
    127–131
Asian Pacific Environmental
    Network, 24, 175–176
Association of People for Practical
    Life Education, Ghana, 14
Atlanta Constitution, 20

Baker, John S., Jr., 41
BANANA, defined, 243–244
Barrio, defined, 244
Basel Convention on the Control
    of Transboundary

Movements of Hazardous
    Wastes and Their Disposal,
    49, 62, 131–134
Baucus, Max, 64
Been, Vicki, 42
Benign neglect, defined, 244
Best available technology (BAT),
    defined, 244
"Big Ten", 63
Biodiversity, defined, 244
Birth defect, defined, 244
Black Congressional Caucus, de-
    fined, 244
Blaming the victim, defined, 244
Blue-collar workers, defined, 244
Boerner, Christopher, 46
Boomerang effect, defined, 244
Boycott, defined, 244
Browner, Carol, 68
Bryant, Bunyan, 22, 29, 31, 34, 68,
    159
Bryant, Pat, 85, 155
Buffer zone, defined, 244
Bulkhandling, Inc., 47
Bullard, Robert, 5, 6, 7, 18, 19, 21,
    23, 33, 36, 39, 42, 45, 69
Burden of proof, defined, 245
Bush, George, 22
Business-as-usual, defined, 245
Buyout, defined, 245

California Indians for Cultural
    and Environmental
    Protection (Santa Ysabel,
    California), 24
Canada Alliance in Solidarity
    with Native Peoples, 176–177

"Cancer Alley", 9–11, 6
  defined, 245
Capability analysis, defined, 245
Carbamate, defined, 245
Carcinogen, defined, 245
Carson, Rachel, 16
Center for Policy Alternatives,
  177–178
Center for Third World
  Organizing, 178
Center Springs, Louisiana, 10
Centro de Informacion,
  Investigacion, y Educacion
  Social, 178
Chávez, César, 70–71
Chavis, Benjamin, 4, 23, 39, 71–72,
  159
Chemical Waste Management,
  Inc. (CWM), 30, 35
Children's Environmental Health
  Network, 179
Chlorinated hydrocarbon, de-
  fined, 245
Chronic bronchitis, defined, 245
Churchrock, New Mexico, 7–9
Circle of poison, defined, 246
Citizens Clearinghouse for Haz-
  ardous Waste, 17, 179–180
Citizens Coal Council, 180
Citizens' Environmental
  Coalition, 180–181
Citizens for a Better America, 83,
  181
Citizens for a Better Environment
  (CBE), 209
Citizens for Farm Labor, 82
Citizens League Opposed to
  Unwanted Toxins (Tifton,
  Georgia), 24
City Care: A Conference on the
  Urban Environment, 60
Civil disobedience, defined, 246
Civil rights, defined, 246
Civil Rights Act of 1875, 57
Civil Rights Act of 1964, 19, 58,
  88–91
Civil rights movement, 1, 2, 18–19
Clark Atlanta Environmental
  Justice Resource Center, 65
Class action suit, defined, 246
Clawback agreement, defined,
  246

Clean Air Act, 59, 60
Clean Sites, Inc., 38
Clean Water Act, 8, 60
  effects on various income
  groups, 37
Cleanup, defined, 246
Clinton, Bill, 23, 64, 65
The Coalition for Environmental
  Consciousness (Ridgeville,
  Alabama), 24
Cohen, Gary, 50
Collins, Cardiss, 64
Commission for Racial Justice of
  the United Church of Christ,
  20–22, 23, 60, 61, 62, 65, 71,
  171–172, 181–182, 214, 218
Communities of color, 3
  defined, 246
Community buyout, defined, 246
Community information state-
  ment, defined, 246
Community right-to-know, de-
  fined, 246
Community Service Organization
  (CSO), 70
Comprehensive Environmental
  Response, Compensation and
  Liability Act of 1980, 105–109
Concerned Citizens of Sunland
  Park (Sunland Park, New
  Mexico), 24
Conference on Race and the
  Incidence of Environmental
  Hazards, 68
Conservation, 15
Corporate Conservation Council,
  National Wildlife Federation,
  26
Correlation, defined, 246–247
Cost/benefit analysis, defined,
  247
Cost effectiveness analysis, 247
Council of Energy Resource
  Tribes, 61, 77
Council on Environmental
  Quality, 59
Coyle, Marcia, 38

DAD (decide, announce, defend),
  defined, 247
Dallas Housing Authority, 6
De facto, defined, 247

De jure, defined, 247
Deadly Deception (video), 236
Deafsmith, A Nuclear Folktale
   (video), 236
Debt-for-nature swap, defined,
   247
Deep South Center for
   Environmental Justice, 84,
   182–183
Degradable, defined, 247
Deland, Michael R., 22
Demography, defined, 247
Desert Land Act of 1877, 15
Detroit River, 12–13
Development rights, defined, 247
Discrimination, defined, 247
Discriminatory intent, defined,
   247–248
Disease of adaptation, defined,
   248
Disproportionate, defined, 248
*The Distribution of Outdoor Air
   Pollution by Income and Race:
   1970–1986,* 37–38

Earth Day, 59
Earth Day 1993, 64
Easement, defined, 248
*East Bibb Twiggs Neighborhood
   Association, et al. v. Macon-
   Bibb County Planning &
   Zoning Commission, et al.,*
   137–140
Ecocide, defined, 248
Eco-Justice Working Group
   (National Council of
   Churches), 81
EcoJustice Project, 183
Ecology, defined, 248
EcoNet, 183–184
Economic blackmail, defined, 248
Eco-Rap: Voices from the Hood
   (video), 236–237
Effluent, defined, 248
The Egg: A Journal of Ecojustice, 25
*El Pueblo para el Aire y Agua
   Limpio v. Chemical Waste
   Management,* 149–151
Emancipation Proclamation, 57
Emission standard, defined, 248
Encroachment, defined, 248
Environment, defined, 248

Environmental Action
   Foundation, 17
Environmental Center for New
   Canadians (Toronto,
   Ontario), 24
Environmental discrimination, 4
   defined, 248
Environmental Equal Rights Act
   of 1993, 114–119
Environmental equity, 4
   defined, 249
"Environmental Equity: Reducing
   Risks for All Communities,"
   63
Environmental Equity
   Workgroup (EPA), 62
Environmental high-impact area,
   defined, 249
Environmental impact statement,
   defined, 249
Environmental inequities
   data and statistics, 25–29
   data and statistics; race, and
      economic status, 26–29
   definition, 3
   differences of opinion, 29–31
   examples, 5–15
   international issues, 13–15
   issues of intent, 39–44
   origins, 31–45
   responses to, 45–47
Environmental issues, concerns
   among people of color, 33–34
Environmental justice, 4
   as a social movement, 24–25
   as international issue, 47–49
   defined, 249
   future directions, 49–50
   types, 5
Environmental Justice Act of
   1992, 64, 79
*Environmental Justice and
   Transportation: Building Model
   Partnerships,* 65
Environmental justice movement,
   history, 19–24
Environmental Justice Resource
   Center, 184
Environmental Protection Agency
   (EPA), 59
*Environmental Quality and Social
   Justice in Urban America,* 17

Environmental racism, 4, 39–44
  defined, 249
Environmental regulations, effects
  on communities of color and
  poor people, 36–39
Environmental Research
  Foundation, 185
Environmental screening, de-
  fined, 249
Environmental Services Office of
  the Cherokee Nation
  (Oklahoma), 24
Environmental Support Center, 81
Environmentalist movement, his-
  tory, 15–18
Environmentally disadvantaged
  community, defined, 249
Environmentally sound manage-
  ment, defined, 249
"Equity in Environmental Health:
  Research Issues and Needs,"
  64
Ethnic, defined, 249
Eufaula Street Landfill Committee
  (Fayetteville, North
  Carolina), 24
Everyone's Backyard, 25
Executive order, defined, 249
Executive Order 11514, 95–97
Executive Order 12898, 23, 65, 74,
  119–127
External cost, defined, 249

Fair Housing Act of 1968, 19, 59,
  91–92
Fair market value, defined,
  249–250
"Fair share" act (New York City),
  63
Fair share legislation, defined, 250
Fauntroy, Walter E., 19
Federal Insecticide, Fungicide,
  and Rodenticide Act of 1947,
  58, 62
  Amendments of 1975, 59
Federal Interagency Working
  Group on Environmental
  Justice, 65
Federal Water Pollution Control
  Act, 8
Feedlot, defined, 250
Feldman, Jay, 72–73

Ferris, Deeohn, 73–74
First National People of Color
  Environmental Leadership
  Summit, 23–24, 63, 68, 81, 82,
  84
Flint-Genessee United for Action,
  Justice, and Environmental
  Safety (Flint, Michigan), 24
Forest Grove, Louisiana, 10
Fort Greene Community Action
  Network (Brooklyn, New
  York), 24
Fossil fuel, defined, 250
Friends of the Earth, 17
*From the Mountains to the
  Maquiladoras: A TIRN
  Educational Video* (video), 237

Gammalin, 20, 13–14
García Martínez, Neftalí, 74
Gaylord, Clarice E., 75
Gelobter, Michel, 37–38, 159
General Accounting Office, 42, 61
Genetic damage, defined, 250
Gentrification, defined, 250
Geographic equity, 5
Get It Together (video), 237–238
Gianessi, Leonard P., 37
Global Dumping Ground (video),
  238
Godsil, Rachel D., 63
Goldman, Benjamin A., 26–29,
  43–44
Goldtooth, Tom, 85
Gore, Al, 64
Grandfather rights, defined, 250
Grassroots, defined, 250
Great Louisiana Toxic March, 62
Green revolution, defined, 250
*Greenbucks: The Challenge of
  Sustainable Development*
  (video), 238–239
Groundwater, defined, 250
Growth management, defined,
  250
GSX Corporation, 45
Gulf Coast Tenants Organization,
  24, 62, 185
Gutierrez, Juana Beatriz, 75–76

Hahn-Baker, David, 159
Hamilton, James T., 43

Harris, LaDonna, 76–77
Harrisburg Coalition against
Ruining the Environment v.
Volpe, 146–148
Harrison, David, 21
Hart, Philip, 59
Harvard Center for Risk Analysis,
185–186
Harvest of Sorrow: Farm Workers
and Pesticides (video), 239
Hatcher, Richard, 18
Hazardous waste, defined, 250
Heavy metal, defined, 250–251
Herbicide, defined, 251
Heroes of the Earth (video), 240
Highlander Education and
Research Center, 186–187
Hispanic American
defined, 251
exposure to pesticides, 12
Historic preservation, defined,
251
Historically black colleges and
universities (HBCU), defined,
251
Host fee, defined, 251
Hot potato, defined, 251
Hot spot, defined, 251
Houston, Texas, 21

Impact analysis, defined, 251
Incineration, defined, 251
Indian Law Resource Center,
187–188
Indigenous people, defined, 251
Industrial smog, defined, 251
Insecticide, defined, 251
Institute for Local Self-Reliance,
188
Institutional racism, defined, 251
Integrated pest management
(IPM), defined, 251
Intent (legal) in environmental in-
equities, 44–45
Interagency Symposium on
Health Research and Needs
to Ensure Environmental
Justice, 65
Interagency Working Group on
Environmental Justice, 65
Internal costs, defined, 252

Jeffreys, Kent, 40
Job blackmail, 35–36
defined, 252
Johnson, Gary, 31
Johnson, Hazel, 85

Kassa Island (Guinea), 47, 62
Ke Kua'aina Hanauna Hou
(Kaunakakai, Hawaii), 24
Kerr-McGee Nuclear Corporation,
8
Koko, Nigeria, 61
Krauss, Celene, 25

Labor Institute, 78, 188–189
Labor Occupational Health
Program (LOHP), 78, 189–190
Laissez-faire, defined, 252
Lambert, Thomas, 46
Land use, defined, 252
Landfill, defined, 252
Latino(a), defined, 252
Lavelle, Marianne, 38
Lawyer's Committee for Civil
Rights Under Law, 73
LD-50 (lethal dosage-50 percent),
defined, 252
Lee, Charles, 85, 159
Lee, Pamela Tau, 77–78
Leopold, Les, 78
Less-developed countries (LDC),
defined, 252
Lewis, John, 64, 79
Lincoln, Abraham, 57
Lindane, 14
Litigation, defined, 252
Living with Lead (video), 240
Lobby, defined, 252
Locally unwanted land uses, 33,
35, 40–47
defined, 252
Louisiana Advisory Committee to
the U.S. Commission on Civil
Rights, 10, 30, 41
Louisiana Department of
Environmental Quality, 31
Low-income housing, defined,
252
LULU. *See* Locally unwanted land
uses
Lumbee Indians, 45

Madres del Este de Los Angeles
(Mothers of East Los
Angeles; MELA), 62, 75, 190
Mainstream environmentalism, 17
defined, 252
Maquiladoras, defined, 252
March on Selma, 1965, 19
March on Washington, 1963, 19
*Margaret Bean et al. v. Southwestern
Waste Management Corp. et al.*,
134–137
Market forces, defined, 253
Mexican Americans, exposure to
pesticides, 12
Michigan Coalition, 22, 62, 84
Migrant Legal Action Program,
Inc., 190–191
Miller, Vernice, 85
Minority Environmental Health
Conference, 63
Model Environmental Justice Act,
162–171
Mohai, Paul, 22, 29, 31, 34, 79–80,
159
Monitored Retrieval Storage
(M.R.S.), defined, 253
*The Moon's Prayer: Wisdom of the
Ages* (video), 240–241
Moore, Richard, 81
More-developed countries
(MDC), defined, 253
Morrisonville, Louisiana, 10
Moses, Marion, 81–82
Mothers of East Los Angeles. *See*
Madres del Este de Los
Angeles
Move-to-the-nuisance, defined,
253
Muir, John, 16
Multiple use, sustained yield,
15
Murph Metals secondary lead
smelter, 6
Mutagen, defined, 253

National American Indian
Environmental Illness
Foundation, 191
National Association for the
Advancement of Colored
People, 72
National Coalition Against the

Misuse of Pesticides
(NCAMP), 72, 191–192
National Conference of Black
Lawyers, 192–193
National Congress of American
Indians, 24, 193–194
National Environmental Justice
Advisory Council (NEJAC),
23, 64, 68, 74, 81, 84
National Environmental Policy
Act of 1969, 59, 92–95
National Farm Workers
Association (NFWA), 70
National Indian Business
Association, 77
National Indian Housing Council,
77
National Institute of
Environmental Health
Sciences, 64, 194
National Law Journal, 63
National Parks and Conservation
Association, 16
National Tribal Council on the
Environment, 77
National Urban League, 60
National Women's Political
Caucus, 77
Native Americans for a Clean
Environment, 194–195
Natural Resources Defense
Council, 195–196
Navajo Nation, 7–9
Negative externality, defined, 253
Negotiated compensation, de-
fined, 253
Networking, defined, 253
New Federalism, defined, 253
New York City, 63
NIMBY. *See* "Not in my back-
yard"
nitrogen oxides, defined, 253
Nixon, Richard, 59
Nonconforming use, defined, 254
Nongovernmental organization,
defined, 254
Nonviolence, 19
NOPE (No place on Earth), de-
fined, 254
"Not in my backyard" (NIMBY),
32–33
defined, 253

Not Just Prosperity: Achieving Sustainability with Environmental Justice, 26
Nuclear waste, defined, 254
Nuisance law, defined, 254

Office of Environmental Justice (EPA), 23, 63, 75
Offsetting benefits, defined, 254
Oklahomans for Indian Opportunity, 76
Open dump, defined, 254

Panna Outlook, 25
The Panos Institute, 196
Parks, Rosa, 2, 18
Particulate matter, defined, 254
People Concerned about MIC [methyl isocyanate], 61
People for Community Recovery, 196–197
People of color, defined, 254
People Organized in Defense of Earth and its Resources (Texas), 24
Persistent, defined, 254
Peskin, Henry M., 37
Pesticide, defined, 254–255
Pesticide Education Center, 82, 197
Pesticide exposure, 11–12
Pesticide treadmill, defined, 255
Petrochemical, defined, 255
Petrochemical colonialism, defined, 255
pH, defined, 255
Philadelphia, waste disposal policies, 47, 61, 62
Photochemical smog, defined, 255
PIBBY. *See* "Put it in the blacks' backyard"
Pinchot, Gifford, 16
Plessy v. Ferguson, 57
Point source, defined, 255
Pollution, defined, 255
Polychlorinated biphenyls (PCBs), 1–2
defined, 255
Porter, J. Winston, 39–40
Preemption Act of 1841, 15
Preservation, 16
Presidential Transition Team, U.S.

Environmental Protection Agency, recommendations, 23, 151–155
"Principles of Environmental Justice," 23–24, 63, 156–158
Prior informed consent, defined, 255
Probst, Kate, 38
Procedural equity, 5
Project Que: Environmental Concerns in the Inner City, 60
Proof of intent, defined, 255
Property value, defined, 255
Puerco Valley Navajo Clean Water Association, 9
"Put it in the blacks' backyard" (PIBBY), 32–33
defined, 255

Quality of life, defined, 256

"Race and the Incidence of Environmental Hazards" (Michigan Conference), 22, 62, 158–162
Race, Poverty, and the Environment, 25
RACHEL's Hazardous Waste News, 25
Racism, 5
defined, 256
Radioactive waste, defined, 256
Rebuttable presumption, defined, 256
Recycling, defined, 256
Refuse Act of 1899, 58
Reilly, William K., 22, 68, 158
Reserved Rights Doctrine, defined, 256
Resource Conservation and Recovery Act, 60, 109–112
Resource recovery, 256
Revilletown, Louisiana, 10
Right-to-inspect, defined, 256
Right-to-know, defined, 256
Rio Puerco Valley, 7–9
R.I.S.E., Inc., et al. v. Robert A. Kay, Jr., et al., 140–141
Risk assessment, defined, 256
The River that Harms (video), 241
Robinson, William Paul, 8–9
Roosevelt, Theodore, 16

RSR lead smelter, 6–7
Rule 1057 (State of Michigan), 12
Runoff, defined, 256–257

Sacrifice zone, defined, 257
Saika, Peggy, 85
Sanitary landfill, defined, 257
Santa Fe Health Education
    Project, 197–198
Secured landfill, defined, 257
Self-determination, defined, 257
Servicios Cientificos y Tecnicos,
    74
Shepard, Peggy, 82–83
Shocco Township, North
    Carolina, 1–2
Sierra Club, 16, 17, 60
Sierra Club Legal Defense Fund,
    198–199
Silent Spring, The, 16
Site, defined, 257
Sit-in, defined, 257
Small, Gail, 85
Smith, James Noel, 17
Social equity, 5
Solid Waste Disposal Act of 1965,
    101–105
Source reduction, defined, 257
Southern Organizing Committee
    for Economic and Social
    Justice, 62
Southwest Network for
    Environmental and Economic
    Justice, 24, 81, 160–161,
    199–200
Southwest Organizing Project, 62,
    81, 200
Southwest Research and
    Information Center, 8
Standard Metropolitan Statistical
    Area (SMSA), defined, 257
Stressors, defined, 257
Strip mining, defined, 257
Subsidence, defined, 257
Suitability analysis, defined, 258
Sulfur oxides, defined, 258
Sullivan, Louis W., 22
Summers, Lawrence, 48
Sumter County, Alabama, 35
Sunrise, Louisiana, 10
Sunshine laws, defined, 258
Superfund, defined, 258

Super-pests, defined, 258
Surface mining, defined, 258
Sustainable development, de-
    fined, 258
Symposium for Health Research
    and Needs to Ensure
    Environmental Justice, 68
"Synergy '94: The Community
    Responsibilities of Sustain-
    able Development," 26, 217

"Taking Back Our Health — An
    Institute on Surviving the
    Toxic Threat to Minority
    Communities," 60
Task Force on Environmental
    Equity and Justice (Texas), 64
Teratogen, defined, 258
Texas Air Control Board, 64
Texas Water Commission, 64
Third World, defined, 258
Thoreau, Henry David, 16
Threshold effect, defined, 258
Throw-away community, defined,
    258
Tools of protest, available to peo-
    ple of color, 34
Tort, defined, 259
Toxic, defined, 259
Toxic colonialism, defined, 259
Toxic fish, consumption, 12–13
Toxic Racism (video), 241
Toxic Substances Control Act of
    1976, 60, 112–114
Toxic Times, 25
"Toxic Wastes and Race in the
    United States," 61
"Toxic Waste and Race Revisited,"
    65
Toxin, defined, 259
Tucker, Connie, 85
Tucker, Cora, 83–84
Twenty-fourth Amendment, 58

"Unequal Environmental
    Protection," 63
United Auto Workers, 60
United Farm Workers of America,
    AFL-CIO, 71, 82, 200–201
United Nuclear Corporation, 8
U.S. Department of Transpor-
    tation, 65

U.S. Environmental Protection
Agency (EPA), 2, 63
enforcement policies, 38
Office of Environmental Justice,
201–202
Statutory Civil Rights
Requirements, 98–101
U.S. General Accounting Office,
20, 230
U.S. Supreme Court, 58, 61, 134
Urban Environment Conference,
60
Urban League, 60

Variance, defined, 259
The Video Project, 202–203
*Village of Arlington Heights et al. v.
Metropolitan Housing Develop-
ment Corp. et al.*, 144–146
Voces Unidas, 25
(To) vote with one's feet, defined,
259
Voting Rights Act of 1957, 19, 58
Voting Rights Act of 1965, 59

Wallace, Louisiana, 10
Ward Transformer Company, 1
Warren County, North Carolina,
1–3, 19, 60, 224
Washington, Mayor of Washington,

D.C., et al. v. Davis et al., 44,
142–144
Washington Office on
Environmental Justice, 74,
203
Water Pollution Control Act of
1948, 58
Water Pollution Control Act of
1972, 59
Water table, defined, 259
West Dallas, Texas, 6–7
West Harlem Environmental
ACTion (WHE ACT), 83,
203–204
West, Patrick C., 12–13
White flight, defined, 259
Wilderness Society, 16
Willow Springs, Louisiana, 10
WMX Technologies, 29, 30
Wolff, Edward, 37
Women, role in environmental
justice movement, 25
Working Group on Community
Right-to-Know, 204–205
Wright, Beverly, 84–85, 159

The Yes! Tour: Working for Change
(video), 242

Zoning, defined, 259

D avid E. Newton holds B.A. and M.A. degrees from the University of Michigan and an Ed.D. in science education from Harvard University. He taught mathematics and science at the secondary level in his hometown of Grand Rapids, Michigan, and then courses in chemistry, physical science, teacher education, and human sexuality at Salem State College in Salem, Massachusetts. He has also held appointments as visiting professor at Western Washington University and as adjunct professor at the University of San Francisco. He has more than 400 publications to his credit, including 50 books on topics such as science and social issues, gun control, hunting, global warming, the gay and lesbian civil rights movement, ozone depletion, and the chemical elements. He and his partner own a nine-room bed and breakfast inn in Ashland, Oregon.